ENERGY, ECONOMIC GROWTH, AND REGIONALISM IN THE WEST

ENERGY, ECONOMIC GROWTH, AND REGIONALISM IN THE WEST

Lynton R. Hayes

Foreword by
Richard D. Lamm, Governor of Colorado

UNIVERSITY OF NEW MEXICO PRESS
Albuquerque

Library of Congress Cataloging in Publication Data

Hayes, Lynton R. 1943—
Energy, economic growth, and regionalism in the West.

Includes bibliographical references and index.
1. Energy policy—The West. 2. Power resources—
The West. 3. The West—Economic conditions.
4. Western Governors Policy Office. I. Title.
HD9502.U53A1714 333.7 79-4932
ISBN 0-8263-0515-6

 Manufactured in the United States of America. Library of Congress Catalog Card Number 79-4932. International Standard Book Number 0-8263-0515-6.

First edition

I dedicate this book
to my exceptional partner, Shannon,
and to our wonderful small children,
Fleur, Courtney, and Joanna.
We offer thanks to the Hayes and Byrne families
in Western Australia
for their support in our studies abroad.

Acknowledgments

As the original sponsor of my education in the United States, Lynton K. Caldwell has given me encouragement, guidance, and opportunities to progress seldom provided for a student. I will always be grateful for his contributions to this study.

Without the financial support of the Natural and Environmental Sciences Program of the Rockefeller Foundation, this book could not have been completed. A special debt of gratitude is owed to Dr. Ralph W. Richardson, Jr., and Dr. Gary W. Toenniessen for their encouragement and support.

Many people in the West gave generously of their time and energy for this cause. Special thanks go to the staff of WGREPO and WESTPO, especially to Dr. Philip Burgess, who provided stimulation, insights, and essential documents beyond the call of duty. Supporters who at one time or another examined the manuscript or helped in its preparation include John Watson of WIEB/WINB, Beth Hadas of the University of New Mexico Press, and Edna Colby, Muriel Keeler, Priscilla Myers, and Jane Marques of Pensacola. To others in the West, and in Washington, D.C., who contributed and cannot be listed individually I hope the book is a measure of my gratitude. I reserve for myself responsibility for all the work's shortcomings.

Contents

Acknowledgments vi
Foreword, by Richard D. Lamm, Governor of Colorado ix
List of Acronyms and Abbreviations xii
1 The Rocky Mountain and Northern Plains Region: Institutional Response to Rapid Change 1
2 Energy Resources in the American West: Development Programs and Anticipated Impacts, 1970–2000 11
3 Toward Understanding the American West and Its Experiences as a Region 51
4 Emergent Conflict over Energy Policies: Region Versus Nation? 75
5 Organizing for Gubernatorial Effectiveness: Energy Development as a Catalyst 123
6 The Institutional Response of a Region: An Interpretation 163
Appendix: Regionalism and Regional Institutions as Subjects for Political Analysis 191
Notes 219
Index 233

Figures

1. General Distribution of Energy Resources in Eight Western States 14
2. Organization of the Western Governors' Policy Council 143
3. Organization of the Western Governors' Policy Office 180

Table

1. Expectations of Energy Facility Expansion 38

Foreword

New Institutions for National Energy Policy

Richard D. Lamm, Governor of Colorado

Lynton R. Hayes's *Energy, Economic Growth, and Regionalism in the West* is a timely account of how the energy-producing Western states have attempted to work together through new political institutions to help meet the nation's energy needs while protecting environmental and other values of the West that are equally important to present and future generations of Americans.

Professor Hayes writes insightfully about an important dimension of national energy policy: the institutional dimension. This is not a typical approach. Many policymakers, experts, and commentators carry on about *supply* uncertainties—for example, the fact that much of our fuel supply is controlled by foreign sovereigns and can be reduced or cut off at will (as in the Arab boycott of 1973) or as a consequence of political instability (as in the coup in Iran in 1979). Others emphasize *demand* uncertainties—for example, the fact that we know so little about how to make demand-reducing conservation policies and programs work. Still others focus on *performance* uncertainties, such as the fact that most of our alternative energy sources (synfuels, geothermal, solar, and others) are based to a large extent on state-of-the-art technologies where successful commercialization depends largely on scientific or technical breakthroughs which are difficult, at best, to anticipate. Unlike these more typical approaches, Hayes shows that we need to give more attention to *institutional* uncertainties—e.g., by our continuing institutional incapacity to facilitate intergovernmental policy and program development among federal, state, and local governments and our inability to invent institutions that effectively link the public and private sectors.

As Professor Hayes points out, the governors and other leaders of Alaska and the intermountain West have given considerable time and attention to shaping new institutions to help us meet new challenges and opportunities in energy and related areas, including water and transportation.

As new challenges and opportunities emerge, so must our institutions. In the ten-state WESTPO area (Alaska, Arizona, Colorado, Montana, Nebraska, New Mexico, North Dakota, South Dakota, Utah, and Wyoming), successive governors have shared a strong conviction that we must work together if we are to be heard in a political system that, to a large extent, measures political influence in terms of population. In the sparsely populated WESTPO states—with nearly one-fourth of the nation's land area and less than 5 percent of the population—we cannot go it alone. The Western spirit of independence, while laudatory, can work against our long-term interests. Individually we can be ignored. Collectively we can play a significant and constructive role in shaping national policy and in helping to meet national needs while preserving the unique qualities of the intermountain West. The evolution of WESTPO, described and chronicled by Professor Hayes, represents a constructive approach to regionalism and the role of states in shaping national policy, including federal policy.

Professor Hayes clearly outlines the common problems and opportunities of the West—aridity, vast and sparsely populated areas, and rich mineral endowments. He also cites the common geography, including rugged mountains and rolling plains, to help account for our successes in shaping new institutional arrangements.

In addition to these and other factors, I would like to underscore the fact that the Western governors who have worked together on these issues since the 1973–74 oil embargo have built relationships firmly based on mutual affection and respect. This has permitted us to try novel approaches to regional problems in an atmosphere of great trust, high tolerance, and mutual confidence. These are important ingredients for success in any process of institution building, but especially among sovereign states.

The efforts of the Western governors, through WESTPO and other institutions, to deal more effectively with regional issues and with the federal government represent significant milestones in the history of intergovernmental relations. Finding ways and means to strengthen our states and the multistate institutions created to serve state interests in the intergovernmental system is indeed a major national need, genuinely felt among the people of the Western states. Professor Hayes's research reveals both the excitement and the frustration of our search for more effective and responsive institutions of government—and the hopes many of us share to strengthen the role of the states in the American federal system and to improve thereby our capacity to conduct the public's business.

List of Acronyms and Abbreviations

EPA	U.S. Environmental Protection Agency
ERDA	U.S. Energy Research and Development Administration
FEA	Federal Energy Administration
Federation	The Federation of Rocky Mountain States
Four Corners	Four Corners Regional Commission
MSO	Multistate organization
NGPRP	Northern Great Plains Resource Program
Old West	Old West Regional Commission
Pacific North West	Pacific North West Regional Commission
REAP	Regional Environmental Assessment Program
Task Force	Western Governors' Task Force on Regional Policy Management
WASDA	Western Association of State Departments of Agriculture
WFHS	Western Federation for Human Services
WESTPO	Western Governors' Policy Office
WGC	Western Governors' Conference
WGREPO	Western Governors' Regional Energy Policy Office
WICHE	Western Interstate Commission on Higher Education
WIEB/WINB	Western Interstate Energy Board/Western Interstate Nuclear Board
WINB	Western Interstate Nuclear Board
WSWC	Western States' Water Council

1 The Rocky Mountain and Northern Plains Region: Institutional Response to Rapid Change

As recently as the 1960s, the states of the Rocky Mountain West and Northern Great Plains were striving to attract economic and industrial development. The objective of economic growth was common to all these states, but as teams of industrial promoters from the states combed the financial and business capitals of the nation attempting to attract industry and people, interstate rivalry resulted. At the time, development of the vast fossil fuel deposits which lay buried in many of these states would probably have been welcomed, but there was no market for the energy. Then the market developed. The nation recognized that its oil and natural gas resources were dwindling rapidly, that its energy-intensive economy demanded a doubling of energy consumption each decade, and that low-sulfur coal was needed to meet air quality standards in the cities. Suddenly, the West's vast low-sulfur strippable coal resources, oil shale, uranium, and geothermal fuels appeared as essential to the nation; surely they would stimulate the economy of the region. By

1970 major energy development programs had been designed by industries, utilities, and federal agencies to extract, generate, and distribute energy in Montana, North Dakota, Wyoming, Colorado, Utah, Arizona, and New Mexico. By far the greatest proportion of the power was to be exported to West Coast, Northwest, and Midwest consumers. Today the region is the nation's most richly endowed energy frontier.

Developing energy resources has environmental consequences at each stage of the process—from exploration to the disposal of waste residuals. Landscape degradation, several forms of air and water pollution, weather modification, preemption of vast quantities of water, disturbance of wildlife habitats, increase in noise, and the changes in life-style that accompany the spread of settlements and large population increases are each considered as part of the environmental impact. Each of the aforementioned Western states is today experiencing certain of these impacts. Neither atmospheric systems nor hydrological systems, however, are neatly contained within state borders, nor are the effects of widespread industrialization and urbanization. Many of the environmental problems related to energy development in the West are felt region-wide. Neighboring states such as Nebraska, Nevada, and South Dakota, which are not endowed with significant energy resources, are affected as well. Environmental problems in contiguous states are often so shared that solutions to them would appear to require a multistate or regional approach. Not only do the problems of environmental management and quality of life extend throughout the region, but also federal energy policy for the vast area of energy-rich public lands within most of the states has regionwide implications.

In brief, many of the policy issues begun in the 1960s are regionally pervasive. Summarily, these issues relate to extraordinary population growth, increased competition for already scarce water, diversion of water from agriculture to industry, quality of life issues, energy development policies, and the conflict and cooperation between state and federal governments over policy preferences and preemptory rights. Together with incidental problems, these issues have, especially since about 1972, created substantial political contention in the Western states.

Within a brief period environmental, land preservation, and Indian groups have become well-organized and forceful. State legislation aimed at preserving the quality of life in the West and at controlling major aspects of industrial activity has proliferated. The period has also seen, for the first time in the history of the West, Democratic gubernatorial rule among the contiguous energy-rich states and their neighbors. A few governors were elected on platforms of "growth control" and the "preservation of Western life-style" in the face of one of the nation's most massive industrial development drives. Public announcements attest to gubernatorial assertiveness in these states, providing a background for political and institutional innovations.

Governor Thomas Judge of Montana spoke out in October 1974, arguing that "decisions should be made by the people of this region, not by the federal government or the energy companies. If we sit passively by, we will be a colony for the rest of the United States."[1] "We've got to organize, or the Eastern banks and corporations, helped by the Ford Administration, will pick us off one by one."[2] For months prior to this time, Judge had been striving to create a regional alliance to control energy and resource development in the West. This philosophy was well supported by former Lieutenant Governor Bill Christiansen, also of Montana, who earlier in the same year stated, "Montana does not intend to become a boiler room for the nation."[3]

Soon after his election to office late in 1974, Governor Richard Lamm of Colorado pledged that the West would fight indiscriminate development by energy companies: "We won't become the nation's slag heap. I am convinced my election is due in part to a new assertiveness. There is a whole new breed of people out here."[4] At the Democratic Governors' Conference in Hilton Head, South Carolina, Lamm, speaking on behalf of the Western states, said: "We have seen what's happened to Kentucky and Tennessee and West Virginia and other states that have been the nation's coal bin, and we're not going to let that happen to us."[5]

In an article entitled "Next 'Civil War' Issue: Energy"—a headline representative of numerous commentaries in Western newspapers during 1974 and 1975—a flamboyant expression of the union developing between the states claimed, "If a second civil war is fought in the United States, it won't be between the North and South, and slavery won't be the issue. Rather it will be between the scenic West and the population-heavy East and the issue will be energy."[6] Governor Ed Herschler of Wyoming, speaking for his state and neighboring states, declared, "We are trying to guarantee . . . that this growth, change and activity will take place on our terms. Anything less than that is unacceptable."[7] In the view of Governor Arthur Link of North Dakota, "Long range agricultural use of land must take precedence over short term energy recovery. . . . a strong agriculture will be the backbone of this area when the last drop of oil and the last lump of coal have been extracted."[8]

At the Western Governors' Conference on Agriculture and Energy held in Billings, Montana, in April 1975, Governor Jerry Apodaca of New Mexico stated: "The West will not become an energy colony for the rest of the nation." During this conference, governors representing ten states brought to climax previous negotiations and meetings held in an effort to institutionalize a united Western front by confirming an earlier announcement made in February in Washington, D.C. It was the formation of the Western Governors' Regional Energy Policy Office. Upon his appointment as director of the new organization, former governor of North Dakota William L. Guy clarified the philosophy underlying the governors' move to unite: "Most of the people in the ten Western states are ahead of the federal government in their demand that a national energy policy be formulated which gives them protection in the quality of life as it relates to energy."[9] "We have reached the end of development at any price of resources in the West," said Guy. "Environmental considerations must prevail over profit making."[10]

Few would deny that the nation's thirst for energy, the accompanying reordering of industrial and demographic concentrations in the Sun Belt states, and the loosely coordinated national policies for addressing such problems pose serious challenges as well as opportunities for the Rocky Mountain region. The states of the region perceive the threat as growing out of the energy problem, and energy issues have consequently triggered the multistate reaction. Of course, the total energy issue is complicated, requiring comprehensive thought.

Broadly defined, the energy issues facing the West are a sociological and political complex. Indeed, there is a hierarchy of issues in focus, especially those of state sovereignty, self-government, regional autonomy, and the need for improving gubernatorial capacities for leadership in state and national policy making. More specifically, the challenges of energy development imply a need for public management of interrelated problems such as industrialization, urban development, land use, water use, agriculture, and more. There is a pressing need to manage a full range of natural resources as well as the scarce human and technical resources thinly spread across the region in a mere 6 percent of the nation's population. Such an undertaking would benefit from harnessing technical resources in the region's industrial sector and the states' universities and joining them with governmental attention to the problems of a region undergoing unprecedented change.

While the accent has been on positive intentions and accomplishments, part of the region's response has been reactionary and somewhat negative. "We'll not be sacrificed and colonized" is the recurring theme of Western governors. A pronounced quasi-isolationist stance existed in Montana and Colorado, especially in the early phase of the development period. On the receiving end of change and of Eastern and federal initiatives, the states reacted defensively and groped for an institutional means to strengthen their political credibility in Washington, D.C. Late in 1974 and early in 1975, it was the consensus among the governors that an aggressive confrontation with the rest of the nation might be necessary. While this spirit of defensive unity has been generally maintained, the cause of confrontation has dwindled in favor of the development of multistate capability in pursuit of an armistice in the second war between the states.

This developing multistate political reaction to the challenges facing the West has focused on an institutional response. Thus far the experiment in intergovernmental relations has centered on gubernatorial initiatives, with the contrasting absence of comparable leadership by Western state legislators or congressional delegations. Gubernatorial initiatives created the former Western Governors' Regional Energy Policy Office (WGREPO) to undertake research, to articulate the views of the states in national energy policy formation, and to bolster the political power of the region against the nation. Subsequently, the governors formed the Western Governors' Policy Office (WESTPO) to extend the work of its predecessor and to increase the capacity of the states to cope with intergovernmental relations. While the principal problems facing WESTPO have to do with synthesizing policy making for energy, water, agriculture, and other vital resources, some of its greatest challenges are posed by competing regional organizations that claim to speak on behalf of the West, consequently fracturing the credibility WESTPO needs in Washington, D.C.

No longer is a slow and belated reaction to Eastern initiatives acceptable to the governors of the region. The states used WGREPO, and more noticeably WESTPO, to build their capacities to participate with the federal government in decision making and perhaps to offer leadership as well. The governors in the WESTPO group have attempted to push well beyond the single-issue focus of WGREPO to encompass the interrelated aspects of the broader energy challenges and to secure their states in the national decision-making process. Nowhere is this more readily apparent than in the first major policy products of WESTPO.[11] At a White House conference on regional economic imbalances, the WESTPO group delivered a policy paper that not only itemized the region's priorities in the complex of energy issues but also took the lead in offering guidelines for national policy making. The West's guidelines proposed new regional institutions for addressing interregional development problems. The case was made for new multistate coalitions to strengthen the role of the states of the American federal system. It would appear that few other regional groups of states are as well prepared to offer advice on altering the course of intergovernmental relations.

Since the early 1970s the Rocky Mountain and Northern Plains states have accumulated more multistate experiences in coping with energy and natural resource development issues than any other contiguous group of states in the nation. These states have replaced the South as the region with the strongest sense of identity and have undertaken one of the most significant experiences in intergovernmental relations since the Civil War. As has been argued cogently elsewhere by Caldwell, Kash, and others, the nation's energy problems act as catalysts and pose important institutional challenges to the federal system.[12] An examination of the West's recent innovations is instructive in part because they represent a major effort to strengthen the gubernatorial policy-making capacity within states, among states, and especially between states and the federal government.

Which are these states, within whose borders the federal government owns an average of 56 percent of the land, and in which much of the impact of Western fossil fuel development is registered? They include Arizona, Colorado, Montana, Nebraska, Nevada, New Mexico, North Dakota, South Dakota, Utah, and Wyoming. Together with Alaska these states have been the principal participants in regional institutional reform for energy policy, with intermittent participation by the Pacific Rim states of Idaho, Washington, Oregon, California, and Hawaii. The ten Rocky Mountain and Northern Plains states have a predominance of public sentiment, bolstered by rugged individualism and populism, that strongly opposes federal intervention and favors self-determination in state policy making. Yet individual attitudes appear to range from being strongly opposed to rapid energy development to strongly supportive, with a general dislike of planning common to all positions. Orderly development in the West is in widespread demand, but opposition to growth and development is less than to notions of colonialism.

Not only is there a shared opposition to federal dominance and intervention, but, with the spectre of the West becoming a national sacrificial area, there are also common interests in preserving states' rights in water management, mining and reclamation regulations, and the control of environmental management. These common interests appear to contribute to a sense of regionalism, but the differences between states in their priorities for energy development and as they compete for scarce water would at the same time inhibit that sense. The early posturing and rhetoric of the governors of Colorado and Montana in 1974 smacked of resource isolationism, parochialism, and perhaps sectionalism. In contrast, four years later these same governors and their neighboring colleagues endorsed a WESTPO plea for an end to separatism and sectionalism between regions. Do the Rocky Mountain and Northern Plains states exhibit regionalism or sectionalism? Is it possible that their institutional actions represent joint state action rather than the regional action they believe they want?

Observers of Western politics and history have pointed out the usefulness of recording the developments of the region in terms of fluctuations and phases of state interaction with the federal government. In this regard, the energy and resource development phase of the 1970s is politically significant. For decades Western issues had been crowded out of the vision of the region's inhabitants by the homogenizing influences of modern industrial society. The sudden rush for the region's fossil fuels has brought the states to a crossroads in their history. A group of governors have deliberately used two regional mechanisms, the WGREPO and WESTPO, to pull Western issues back into focus and to bring them to the attention of their states and the nation. They have in turn stimulated a national debate on regionalism, indirectly instigated coalition politics in other regions, and attempted major institutional innovations. As indicated before, their actions have not been matched in their region by legislators or congressional delegates. As effective regional organization depends on how federal agencies and Congress perceive such an organization—and especially the governors' support for it—there is considerable interest in diagnosing the challenges facing the West and its institutional response.

This book has been organized as follows in order to discover the principal motivations behind the West's actions and to test the extent of regional consciousness. Chapter 2 defines the energy-rich region, the magnitude of development proposals, the intricate connection among energy, economy and environment, and the views of officials in key states on the dimensions of the problems. The problems are arranged in a threefold order of magnitude, each with differing institutional responses. It is here that we see that state involvement in development proposals was minimal, that federal policies gave out conflicting signals, that knowledge was scarce, and that the institutional setting within which Western development is proposed was confused.

Chapter 3 is a short account of the West as a region, its characteristic features, its previous experience with regional and sectional movements, and the political and legislative initiatives that spread noticeably during the 1970s. Differences in the objectives and public preferences of the states are vast. Although there has never been a political West, multistate cooperation and executive branch solidarity have been growing, particularly in the northern tier of states. The common foe has been an unresponsive and irresponsible federal government.

Chapter 4 presents the politics of creation of the WGREPO—the actions, the process, and the contention that the prime motivation for a multistate bloc is to seek bargaining power relative to federal agencies and national policy makers. The governors pursued impact assistance funding and tried to entice federal energy research and development funds to the West and to defend their rights as states in matters of water use, energy facility siting, and environmental management.

Chapter 5 presents the politics of creation of WESTPO, and the argument that it was designed to continue WGREPO work within a larger context. The push for increased executive branch capacity and strength in dealing with crosscutting problems was immersed in the pursuit of control over other proliferating multistate organizations and the national trend of reasserting the role of the states in the federal system.

In the final chapter, conclusions are drawn about the response of the West to the prospect of rapid development. It is here that the inducers and inhibitors of regionalism, and the views of opinion leaders in the West and the East on the subject of Western organizational initiatives are examined. An appendix provides a brief account of the nature of the concepts of regionalism and sectionalism.

2 Energy Resources in the American West: Development Programs and Anticipated Impacts, 1970–2000

This chapter examines the energy resources in the region, the major development programs announced to date, and the major anticipated environmental and social impacts of development during the next quarter-century. The impacts of energy development—the problems and the benefits—will differ from state to state. Even though each state has substantial reserves of one or another of the fuels of the region, the rate and the magnitude of development that each state can expect are not known either to states or to the federal government. The information at hand does indicate, however, that each state is subject to a wide range of positive and negative effects, from both the supplier and consumer perspectives, if existing development programs are implemented. Two important observations have arisen from this situation. First, each state will have its own prioritized set of energy-related problems. A number of these problems will be uniform to several states, and a number will be peculiar to others. Second, a regional energy policy organization will need to cater to the different problems of different states, the regionally uniform problems, and the vast array of energy-related problems in general.

This chapter will deal with the overall problems of the region, speak to the great variety and interrelatedness of energy impacts, and examine the principal problems identified by each of the states. First, a description of fossil fuels such as coal, oil shale, uranium, oil, and gas includes details of locations, quantity, quality, ownership, and a brief history of development prior to the 1970s. Second is an analysis of the energy development programs, in particular those for coal and oil shale. The analysis is primarily concerned with the origin and objectives of the programs, and a determination of the market to be supplied with the energy. Third, the chapter examines the current impact of development, as well as impacts anticipated from projections of future resource development and growth in the region. Finally, the principal energy-related problems, as seen by energy officials in each of the region's states, are reported. Here energy problems are identified according to whether they can be effectively addressed only with the cooperation of one or more contiguous states (unilateral approach) or whether they can be better addressed by a regionwide multistate approach.

A number of different definitions are used to describe energy resources. Geological research into fossil fuel deposits estimates the amount of resources. Estimates of coal, uranium, and oil resources represent the total stock in the ground according to geological criteria such as the thickness of fuel-bearing strata. Deposits that can be mined using existing technology are known as *reserves*. The part of a reserve that is economically feasible to mine is the *recoverable reserve*.[1] Some recoverable reserves are minable by underground methods; others are available for surface technology or strip mining.

An examination of the energy resources of the region meets with conflicting estimates of resources and reserves. The federal government, state governments, and private owners of the fossil fuels in the region are each unsure of the magnitude of their holdings and of the size of the entire resource deposits. Despite this uncertainty, there is some consensus on the general magnitude of resources and on the quantity and quality of major deposits of reserves. The examination and description that follow focus on the generally accepted estimates for coal, oil shale, uranium, oil, gas, and geothermal fuels. These energy resources are of principal concern because they include the fuels designated for intensive development in the short and medium term.

A general statement of the region's energy storehouse, and one that is often used, is that the Western states, in particular Colorado, North Dakota, South Dakota, Montana, Utah, and Wyoming hold 52 percent of the nation's coal reserves, 48 percent of the uranium, widespread geothermal regions, and significant quantities of oil and natural gas.[2] An even broader statement, made by Governor Richard Lamm of Colorado, claimed the region holds 59 percent of the nation's resources.[3] Despite the generality of such claims, there is no dispute with the position that most of the nation's fossil fuels lie in the Western states, and particularly in the region here considered (see fig. 1). The region's coal resource has been the subject of more rigorous investigation.

West of the Mississippi River lies 57 percent of the nation's coal. The deposits contain 85 percent of the nation's low-sulfur reserves. It has been estimated that 70–77 percent of the nation's strippable coal reserves are in the West. American Indians and the public own 80 percent of these deposits because of concentrations on public lands.[4] Most recent studies indicate that 42 percent of the estimated recoverable reserves of the nation's bituminous and lignite coal are in the ten-state region under investigation in this study.[5]

The coal deposits are not evenly distributed throughout the region. An appreciation of the geographic location of deposits is essential, because development programs, although having regionwide significance, bring about site-specific impacts. Some subregions and localities will feel more coal-related impacts than others, and many have already begun to experience the impacts of development.

The region is composed of two large coal-bearing formations—the Northern Great Plains Province and the Rocky Mountain Province.[6] Each province is composed of coal basins. Within the Northern Great Plains area there are six basins; the Fort Union and Powder River basins are the largest. Within the Rocky Mountain area the Green River, the Uinta, and San Juan River basins are the largest, which, with the Wind River, Bighorn River, Yellowstone, Hams Fork, and Southwestern Utah deposits, make a total of eight basins.

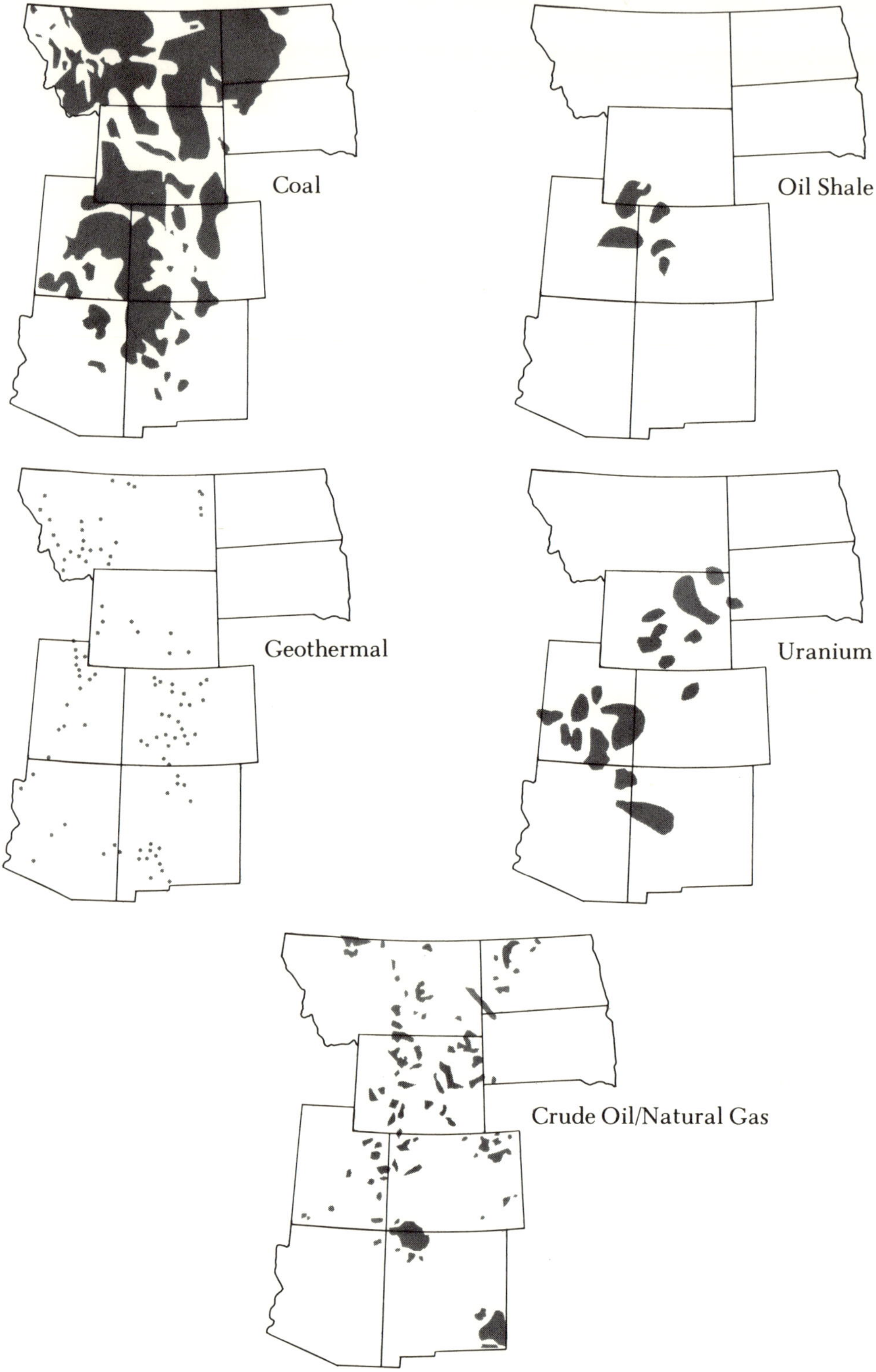

Figure 1. General Distribution of Energy Resources in Eight Western States. Source: *Energy from the West: Draft Policy Analysis Report* (Washington, D.C.: U.S. Environmental Protection Agency, Office of Energy, Minerals, and Industry)

The deposits of the Northern Great Plains Province are concentrated mostly in west-central North Dakota, central-southeastern Montana, northeastern Wyoming, northwestern South Dakota, and east-central Colorado. Here, as in the Rocky Mountain Province, the quality of the coal is a principal determinant of the rate and extent of mining and environmental impacts. Quality may be measured by sulfur content; heating value in British Thermal Units (BTUs), the amount of heat required to raise the temperature of one pound of water one degree Fahrenheit; and the content of ash and moisture.[7] Approximately 90 percent of the coal in the two provinces contains less than 1 percent sulfur. Some experts claim there is a great deal more strippable low sulfur in these provinces than in deposits in the coalfields of the eastern United States, but the ash and moisture content of these western deposits is higher than in eastern coal. Consequently, some argue that the energy potential (BTUs) of the coal in the Central Appalachian low-sulfur deposits exceeds Northern Great Plains coal and therefore weakens arguments for Western coal development.[8] A prominent difference in the quality of coal in the two Western provinces is the BTU range. The Rocky Mountain deposits, bounded by western Montana, west-central Wyoming, east-central Utah, western Colorado, and northwestern New Mexico are of the 6,100-to-13,100-BTU-per-pound range. In contrast, the coal in the Northern Great Plains Province is predominantly in the 6,100-BTU-per-pound range.[9] As briefly mentioned above, the rate and extent of strip mining are determined in part by the low heating value of these deposits and also by the thickness of seams.

Surface mining of the coal seams in the region has been under way for over fifty years, especially in Montana, North Dakota, and Wyoming. Colorado and South Dakota have experienced surface coal mining since the 1930s, and New Mexico has been involved in the activity intensively since the late 1950s. The total production in the region from the 1920s up to early 1972 approximates 300 million tons. This activity disturbed an estimated 20,000 acres.[10] These figures are small fractions of the estimates for future surface tonnage and land disturbance, which will be discussed later in the chapter. Another energy resource that is abundant in the region, oil shale, is also a potential target for intensive mining.

The oil in oil shale is not a free-flowing liquid. The fuel is keragen, trapped in solid form in sedimentary rocks known as *shale*. Oil may be recovered from the shale by heating, which vaporizes it, thus permitting the oil to condense and drain off. Estimates of the quantity of oil shale in the region vary according to the quality of the shale or the productivity expected in terms of barrels per ton. The U.S. Geological Survey estimates 1.82 trillion barrels in rich- and medium-grade shale in the region.[11] This is several times the energy contained in the proven oil and natural gas reserves of the United States. Of the vast oil shale energy resource, some 418 billion barrels have been identified as paramarginal—renderable from ores yielding 25–100 gallons per ton.[12] At present, roughly 100 billion barrels of these reserves are economically and technologically recoverable.

The principal geographic location of these resources is within the Green River formation, which contains 90 percent of all identified resources in the United States. Approximately 80 percent of the Green River shale lands are federally owned. The formation underlies 25,000 square miles, and 17,000 square miles of the ore is believed to contain oil shale with commercial potential. Parts of three states—Colorado, Utah, and Wyoming—occupy these oil shale lands, with Colorado holding 80 percent of the higher grade deposits, Utah holding 15 percent, and Wyoming 5 percent. The richest deposits are in the Piceance Creek basin; four other basins—the Uinta River, Fossil, Washakie, and Sandwash—are also major.

Historically, mining and mineral extraction have been significant parts of the economy in the oil shale belt. Oil and gas, sodium bicarbonate, coal, and some uranium and vanadium have been extracted in the area. The Indians have been recorded as using oil shale for campfires centuries ago. There is no commercial production in the region today, but in 1973 the U.S. Department of the Interior offered six public land tracts for lease in the area. Some estimates have put the productive potential of these six tracts at one million barrels per day by 1985—an amount which would approximate 4 percent of the estimated national demand. While the oil shale resources of the region are an enormous part of the nation's total deposits, its uranium resources are an even greater proportion of the nation's deposits of nuclear fuel.

The states of the region hold about 94.5 percent of the nation's estimated recoverable reserves of uranium.[13] Official U.S. Atomic Energy Commission estimates of discovered reserves, the great part being in the region, today approximate 250,000 tons at a price of eight dollars a pound or 390,000 tons at ten dollars a pound.[14]

Although little information is available for specific subregional operations on production figures or on the quality and quantity of deposits, some significant facts can be observed. Of 191 uranium mines operating in the nation in 1971, all but 11 were in the region. New Mexico holds more than half of the region's estimated recoverable resources; Wyoming, Utah, and Colorado are the other main uranium-producing states with the balance of the reserves. Mining was carried out in North Dakota from 1963 to 1968, has been continued in South Dakota since 1951, and ceased in Montana in 1961. In 1971, the region produced 87.8 percent of the nation's uranium. If the nation continues to move toward rapid escalation of nuclear energy generation, New Mexico and Wyoming, with the greatest share of deposits, will probably be the major sources of future mining.

Whereas the region has the resource base in uranium, oil shale, and coal to make major contributions to national needs, the same is not necessarily the case with petroleum energy. The fuels in this energy group are crude oil, natural gas, and natural gas liquids. Natural gas is the nation's scarest energy resource, and although a *Newsweek* article[15] claimed that the West holds substantial quantities of oil and gas, the facts need closer examination. There was an oil boom in many parts of the region during the 1960s,[16] and the lignite deposits of the Northern Great Plains have good potential for syn-gas production from coal gasification. One recent estimate of the region's petroleum resources is startling for the indicated likelihood of the biggest onshore discovery of oil in the United States in decades.[17] The newly discovered Overthrust Belt, extending through Montana, Idaho, Wyoming, Utah, and western Colorado, has estimated reserves of more than 200 million barrels.

The region is estimated to contain 6.7 percent of the nation's recoverable natural gas.[18] In the northern states of the region, North Dakota, South Dakota, Montana, and Wyoming, reserves were estimated in 1972 at 1,356 million barrels of crude oil,

5,595 billion cubic feet of natural gas, and 1,414 million barrels of natural gas liquids.[19]

Within the region the distribution of petroleum resources is uneven, with New Mexico and Wyoming figuring prominently. These two states appear to be the leading holders of crude oil reserves with 25 percent and 30 percent respectively, but recent reports of the Overthrust Belt reserves may well change conclusions about Colorado, Montana, and Utah having relatively insignificant stocks. The leading holder of natural gas liquids is New Mexico with almost 78 percent; Wyoming has 12 percent of the region's estimated recoverable reserves. With 65 percent and 20 percent, New Mexico and Wyoming also dominate the holdings of the region's natural gas supplies. Montana, Utah, and Colorado each have limited holdings.

Commercial production of petroleum products began decades ago in parts of the region. Wyoming began as far back as 1894, and Montana in 1916. As a combined group, the five Northern Plains states—Montana, North Dakota, South Dakota, Wyoming, and Nebraska—produced a net outflow of 536.9 trillion BTUs of petroleum in 1971, while they imported 37.9 trillion BTUs of natural gas primarily from Canada. Only New Mexico and Wyoming have the potential to export natural gas in the future.

Besides the obvious potential for solar energy generation in the vast, sun-drenched Southwest and Rocky Mountain parts of the region, geothermal energy is thought by many to be a significant energy resource in the West's future. The earth's core is molten rock, or magma, and intense heat radiates from the core to the surface of the globe. Modern drilling technology, reaching depths of a few miles, can reach the geothermal potential claim of about one hundred thousand megawatts. This is roughly equivalent to 25 percent of the total U.S. electric power capacity of 1972. The principal areas of geothermal potential are located in the western volcanic regions in the Pacific and Rocky Mountain states.

Geothermal energy resources are often divided into two classifications. Known geothermal resource areas (KGRAs) form one category, and those classified as "valuable prospectively" the other. As of 1973, estimates put KGRAs at 1.8 million acres, with a further 96 million acres classified as valuable prospectively. Sixty percent of these categories are on public lands,[20] with the bulk of the energy-bearing areas in the region situated in western Colorado, central Wyoming, western Montana, western Utah, and western Idaho. Heightened research and exploration—prompted in part by escalating crude oil prices—together with advances in technology, popular demand for clean fuels with less environmental damage, proposed federal leasing programs, and implementation of the Geothermal Energy Research, Development, and Demonstration Act of 1974 could make the development of geothermal resources feasible from the view of engineering and economics. In Boise, Idaho, for example, homes have been heated by geothermal steam since 1890, and a commercial electric power plant is connected to "The Geysers" in northern California.

The reserves of the various energy resources of the West are immense indeed. The importance of the coal and uranium deposits to the nation in the short and medium term cannot be stressed too heavily, Oil shale and, to a lesser extent, geothermal sources appear of significant value in the medium to longer term if the nation continues to consume at the current rate and if there are not major technological advances in solar power or nuclear fusion. As has been stated, the principal owner of the fossil fuels in the region is the federal government, although the deposits are in certain instances subject to state regulation. Numerous Indian reservations are situated on substantial fuel deposits. The water resources of the region are not plentiful enough for intensive energy development when examined in the context of the historical dominance of agriculture and rigid appropriation rights. These issues will be discussed later in the study, but raising them now forewarns the reader of major problems for the region. To examine thoroughly the energy development implications for the states of the region requires an analysis of the origins, objectives, and scope of the major development programs activated since 1970: Southwest Energy Complex, North Central Power Study, Project Independence, Oil Shale Leasing Program, and President Carter's National Energy Plan.

DEVELOPMENT PROGRAMS

These development programs do not each affect the entire multistate region. Program implementation has taken place in the sequence stated above. The energy development program for the Southwest had its origins in the 1940s, although the basis of several projects in the subregion—massive coal-fired generation—gathered impetus in the middle and late 1960s. As with the North Central Power Study, which was not formulated until 1970–71, the Southwest complex did not create vigorous debate in the affected states until 1971. The U.S. Bureau of Reclamation, in consortium with a number of public and private utilities, was the catalyst in each of these development programs. Rather than being formal government programs, these two projects resembled informal agreements to investigate and to promote opportunities to stimulate economic growth. The Oil Shale Leasing Program in the U.S. Department of Interior, which came into being formally in 1973, was also restricted to a part of the region. Project Independence was announced in 1974, but federal energy rhetoric since that time has not specified which parts of the region would be implicated in the drive toward energy self-sufficiency. Combined, these five programs are expected to continue to bring substantial economic and industrial growth to the region throughout the remainder of the century.

Numerous arguments have been used to present the rationale for the major movement to coal and other fossil fuel deposits in this region of the West, several of which seem most persuasive. Parts of the region have definitely experienced persistent unemployment, low income, and outmigration. The growth orientation of the U.S. Department of Interior, the U.S. Bureau of Reclamation, and industry underlies development programs on public lands and would sustain growth in the area. In part, the drive to utilize Western low-sulfur coal is a response to air-quality standards in urban areas and environmental lawsuits over emissions from, and the siting of, power plants in more populated regions elsewhere. Thus these developments are some of the unintended consequences of the environmental protection movement. Concentrations of energy generation are removed from urban centers to areas in which the plant's polluting and environmentally degrading impacts will have little practical effect.[21] The economics of surface coal mining is also an important consideration. The cost of strip mining coal is about 25 percent of the cost of underground mining.[22] Environmentalists have argued that the motives of utilities in the drive to surface mine Western coal are twofold: to avoid buying scrubbers (equipment to reduce power plant emissions) and to buy low-cost mines from which the companies can profit by obtaining coal from their subsidiaries and by passing costs on to consumers.[23]

One of the most comprehensive statements of possible reasons for the great interest in developing the region's coal was published by the Institute of Ecology; among these reasons were the opportunity to obtain large reserves; lax clean air regulations in some states; thick coal seams close to the surface; relative absence of the United Mine Workers of America; federal ownership of much of the coal; and the scarcity of state and local restrictions on development.[24] It is interesting to note that today state clean air and mining regulations are far from lax in many of these Western states. The motives of the federal agencies for public land and energy development appear to be a mix of the national commitment to growth and efforts since the Arab oil embargo to make the nation energy self-sufficient. The former motive was probably more predominant in the North Central and Southwest development programs, while self-sufficiency was a major goal in the establishment of the Oil Shale Leasing and Project Independence programs.

THE SOUTHWEST ENERGY COMPLEX

The U.S. Bureau of Reclamation has sought to stimulate the economy of the Southwest—New Mexico, Arizona, Utah, and Colorado—for decades. During the 1940s the Central Arizona Project was conceived to implement development plans. In essence these plans were to deliver 1.2 million acre feet of water from the Colorado River into Arizona and western New Mexico by a system of dams, canals, and pumping stations. The Central Arizona Project was eventually authorized in 1968 and the Bridge Canyon and Marble Canyon dams on the Colorado River were proposed as integral parts of the plan. Pressure from conservation and environmental interests prevented the dams from being built, but the U.S. Department of Interior was authorized by Congress to purchase pumping power from nonfederal plants (Public Law 90-537).[25] Also in 1968, the U.S. Department of Interior and Bureau of Reclamation entered into the Navajo

Project Participation Agreement which entitled the Bureau to 24.4 percent ownership of Navajo Power Plant near Page, Arizona.[26]

The Southwest power project is an example of federal sponsorship of a massive development program that was not explicitly authorized by the government of the United States.[27] Before the federal government had become a partner in the Navajo plant agreement, it was involved through the Department of Interior, the Bureau of Reclamation, the Bureau of Land Management, and the Bureau of Indian Affairs in leasing operations with some of the nation's leading energy companies. In 1960 the Peabody Coal Company began exploring and discovered enormous quantities of surface minable coal in the Black Mesa area of northeastern Arizona. These deposits were largely on reservation lands in the Navajo-Hopi Indian Joint Use Area. In 1964 the above government agencies and twenty-three investor and publicly owned utilities formed a consortium known as the Western Energy Supply and Transmission Association (WEST). The WEST consortium planned to construct and operate a coal-fired generating and coal gasification network to produce three times the electric power of the nation's largest utility, the Tennessee Valley Authority.[28]

The electricity produced in the power complex is largely allocated to supply the outlying load centers of Los Angeles, Las Vegas, Phoenix, Tucson, San Diego, and Albuquerque. Population increases in these cities, their astronomical power consumption increases, and a Department of the Interior projection of an increase in population in the Colorado River Basin area from 19.6 million in 1970 to 31 million in 1990 are the bases for expansion of generating capacity. The federal government has based its estimates of electricity in the Southwest at 110 thousand megawatts, of which the coal-fired share will be about 30 percent.[29]

To produce the 30,000–36,000 megawatts of electricity aimed for, the Department of Interior and the private developers saw the need for a four-phase program which would culminate around 1990.[30] In the first phase (1970–77), five main units were built or under construction: the Four Corners plant near Farmington, New Mexico; the Mohave plant in Nevada; the Navajo plant near Page, Arizona; the San Juan plant in New Mexico, and the Huntington Canyon plant in Utah. During the second phase (1978–80), extra capacity was to be added to the existing plants. Kaiparowits, a 5,000–6,000 megawatt facility, once planned for construction in southern Utah, would have been the largest facility of its kind in the world and a major contributor to the combined 13,000 megawatt capacity of all the plants, but it did not survive widespread environmental opposition. Twenty thousand megawatts of power is planned for production by the end of the 1980–85 third phase. The total of 30,000–36,000 megawatts is planned as the total coal-fired capacity in the final 1985–90 phase.

These projects have significant environmental, social, economic, and cultural impacts on both Anglo and Indian inhabitants of this four-state part of the region. In response to demands for review of the need for the programs and a thorough assessment of the environmental and social impacts, Secretary of the Interior Rogers Morton announced in May 1971 the establishment of the Southwest Energy Study.[31] This task-force study was initiated at a time when another energy development complex was being planned in the Northern Great Plains states. The scope of the the latter, the North Central Power Study, was huge, and its productive capacity dwarfed those drawn up for the southern states of the region.

THE NORTH CENTRAL POWER COMPLEX

By 1970 there was general recognition among the energy industry that vast quantities of low-sulfur coal would be needed to fuel the nation's cities where power plant pollution had received public and government disapproval. This recognition, together with the growing awareness of the disadvantages of further dependence on oil from the Middle East, helped turn industrial and governmental attention to exploitation of the American West's coal. As much of the coal lay on public lands, the U.S. Department of Interior began with the belief that major development proposals would require planning and regulation. Jointly, the U.S. Bureau of Reclamation and the thirty-five public and private companies in fourteen states established the North Central Power Study as an informal arrangement with little legal basis.[32] The planners of the energy development laid out a general program that would involve eastern Montana, western Dakota, and Wyoming as producing areas with the capacity to generate more electric power than any country in the world except the United States and the U.S.S.R.[33]

The rush for coal that ensued on the part of industry reminded many Westerners of the rush for gold of the late nineteenth century. Some observers report that buying up large leases of coal as well as options for water to service the plants had begun in 1967 with a water marketing program involving the Yellowstone River system. The power study was initiated in May 1970 by James R. Smith, Assistant Secretary of the Interior for Power and Water Development, and was released in October 1971 under the supervision of Harold Aldrich of the Bureau of Reclamation's regional office in Billings, Montana. The authors of the program worked under the assumption that enormous amounts of power would be required within and outside the area in the future. States to be supplied energy, besides the three in which extensive mining would occur, were Idaho, Utah, Colorado, South Dakota, Nebraska, Kansas, Minnesota, Iowa, and Missouri.

Assuming a growth rate in demand for electricity of 6.5 percent per year, the study participants estimated the need in the market area to be 51,000 megawatts in 1980, and 181,000 megawatts in the year 2000. Of the total electricity production projected from the three-state coal deposits, roughly 20 percent would be allocated to Idaho, Montana, Utah, Wyoming, and Colorado; the balance of 80 percent would go to the Eastern market. In other words, the bulk of the power was for export to markets outside the producing states.

The proposed power complex was to consist of forty-two coal-fired plants. Ten of the plants were to be capable of producing 10,000 megawatts each; that is to say, each plant would have many times the capacity of the massive Four Corners plant in New Mexico. Thirty potential hydro sites were also identified in the study. The number of gasification plants operating by 1985 could reach seven, with a total of sixteen in the three-state area by the end of the century.

The North Central Power Study had originally reflected some desire on behalf of the federal government to guide development in an orderly manner, but the coal and energy companies that were buying up land and purchasing leases and options on coal deposits from the federal government, states, Indians, and private owners rejected the guidelines and engaged in intensive activity in the region. By October 1972, barely one year after the release of the program, opposition from environmentalists, ranchers, Indians, and especially the state of Montana had so escalated that the secretary of the interior took similar steps to the ones he had taken in May 1971 in the Southwest. In response to the chaotic exploitation taking place, the secretary announced the establishment of a joint federal-state task force to assess the environmental, economic, and social impacts in the region—the Northern Great Plains Resource Program (NGPRP). The final reports of the NGPRP were released in August 1975, with examination focused in Montana, Wyoming, North Dakota, South Dakota, and Nebraska, the five states that make up the northern half of the region. A brief examination of the NGPRP findings with regard to anticipated coal development in the five-state area helps to bridge the gap in time between the release of the North Central Power Study and the release of the Blueprint for Project Independence in November 1974.

The NGPRP findings were classified into three scenarios or coal development profiles (CDPs). Each development profile was based on different assumptions about consumption and production. An analysis of the different CDPs permits the reader to estimate what tonnages of coal would be produced in the five states of the subregion through to the year 2000 and what percentages by weight these tonnages would be of national coal production.

The most probable production level chosen by the Energy Work Group of NGPRP was expressed in CDP II.[34] This development profile represents an estimate of the Northern Great Plains coal production rate for export, in-region power generation, electrical export, and synthetic natural gas production based on U.S. Department of Interior forecasts set in 1972.[35] With a production level of 32 million tons in 1973, the subregion produced 5.4 percent of the nation's coal. With an estimated 362-million-ton production in the year 2000, the subregion would produce 25.5 percent of the nation's coal production by weight. The major finding here is that while domestic coal production is to grow by over 100 percent in the next three decades, the contribution of the Northern Great Plains may increase by 1,000 percent.

PROJECT INDEPENDENCE

Federal initiatives in the field of energy planning inched ahead late in 1974 with the release of the Blueprint for Project Independence. The objectives for vastly increased domestic energy production would have major implications for the West. In an energy message on 4 June 1971, President Richard Nixon proposed a program to "ensure an adequate supply of clean energy for the years ahead."[36] The policy of the president in this announcement and through subsequent energy policy moves was to increase supplies from domestic sources and to aim for the highest attainable level of self-sufficiency.

In a televised speech on the energy crisis on 25 November 1973, President Nixon introduced Project Independence.

> What I have called Project Independence-1980 is a series of plans and goals set to insure that by the end of this decade Americans will not have to rely on any source of energy beyond our own. . . . The capacity for self-sufficiency is a great goal.[37]

The tripling of world oil prices in 1973 brought into focus several reasons for increased American energy self-sufficiency. The crippling costs of oil imports created enormous balance-of-payments problems. Imports of oil could cost $50 billion by 1980, whereas $17 billion would have been the cost before the prices tripled. The fear of jeopardizing national security by heavy reliance on foreign oil supplies, together with the global shortages of energy resources, prompted President Nixon's actions to seek a program that would enable the United States to end fuel imports by 1980 without serious damage to the economy.

The program by which the nation would achieve independence required an expansion of 4.7 percent in energy production per year and reduction in the rate of growth in demand for energy from 3.6 percent per year to 2 percent from 1974 to 1980. Supply increase included an oil production hike from 10.9 to 14.0 million barrels per day, a tenfold increase in nuclear power generation, expansion of natural gas production from 23 to

27 trillion cubic feet per year, and conservation savings in households, industry, and transportation. The principal means of supply was to be a substantial coal production increase. The 1973 production of 602 million tons was to increase in 1980 to 962 million tons per year. This would be accomplished by rapid development of coal gasification, setting aside some air quality standards for a five-year period, expansion of coal transportation systems, the conversion of oil burning plants to coal, abandonment of price controls on coal, and a major expansion of the coal mining industry.[38]

The Project Independence Blueprint, begun in March 1974, was designed to provide the foundation for United States energy policy for the future. Prepared by hundreds of federal officials, and dozens of consultants and interagency task forces under the direction of the Federal Energy Administration, the blueprint spelled out the major government actions needed to increase domestic fuel supply to conserve energy and to reduce dependence on foreign fuel. A draft of the blueprint became available in November 1974 and proved to be less ambitious than the program previously outlined by President Nixon.

The Federal Energy Administration concluded in the blueprint that President Nixon's plan to spend $10 billion over five years on energy research and development would not have major impacts on production until the 1985–2000 period. Studies of the feasibility of the president's proposal by the Department of the Interior, the Atomic Energy Commission, and the Energy Policy Project of the Ford Foundation were not optimistic of attaining Project Independence goals even by 1985.[39] Experts calculated that American energy self-sufficiency even by 1985, in contrast to Nixon's 1980, would require doubling coal production, increasing nuclear power capacity fivefold, increasing hydroelectric and geothermal production, increasing oil and gas production 125 percent, and greatly increasing production of synthetic gas from coal and oil from shale. The cost of the plan was estimated at a staggering $600-$900 billion. A report from the National Academy of Engineering estimated that shortages of capital and labor, a dearth of engineering supplies and equipment, and the lack of water in the West for various oil shale, coal, land reclamation, liquefaction, and gasification projects were serious obstacles.

OIL SHALE LEASING PROGRAM

The availability of sufficient water supplies in the West to cope with the proposals of Project Independence, particularly in regard to oil shale mining, generated widespread debate in Washington, D.C., and in the states of the region. The Department of the Interior had attempted to encourage oil shale development by inviting bids for the lease of three oil shale tracts during 1967 and 1968, but the industry rejected the offers. In his 4 June 1971 energy message, President Nixon requested that the Department of the Interior initiate a leasing program, and later that month a draft of the department's Prototype Program was made available. The program involved leasing of six tracts of public oil shale land of approximately 5,000 acres per tract. Two tracts were chosen in each of the states of Colorado, Wyoming, and Utah. The leasing program began in January 1974 with a lease to oil companies of one of the tracts for $210 million. The Department of the Interior estimated that the tracts offered were capable of producing at least one million barrels of oil per day by 1985.

THE NATIONAL ENERGY PLAN (THE CARTER PLAN)

It is difficult to foretell the full impact the Carter Plan will have on Western states, which may well be the cornerstones of the nation's fossil fuel programs. While the Carter Plan emphasizes conservation and a 2 percent reduction in the annual growth of the demand for energy, it also stresses a move to coal.

The movement away from oil and natural gas to coal has wide-ranging implications for the region. Projections that compare 1976 supplies to the coal tonnages needed each year to 1985 indicate that some states must undertake major mining programs. New Mexico's contribution is expected to be increased almost fifteenfold—from 9.4 million to 137 million tons per year; Wyoming's is tenfold, and North Dakota's and Utah's are projected to increase eightfold. At the lower ranges of projected increase are Colorado and Montana, with 400 percent increments, and Arizona, which will not quite double its production.[40]

While the coal production component of the Carter Plan has the most imminent implications for the West, the uranium, oil shale, and solar aspects of the plan are also noteworthy. The reduced emphasis on breeder reactors implies more uranium production in the West, a decision affecting Wyoming and New Mexico especially. Carter's plan has few of the ambitious oil shale proposals pursued by previous administrations, although modest experimentation is being encouraged. The establishment of the Solar Energy Research Institute in Golden, Colorado, offers that state an opportunity to become a principal actor in solar development and perhaps its commercialization. The Department of Energy's forthcoming National Energy Supply Strategy is viewed as the second phase of the Carter Plan, also with important implications for the West. Effort is to be concentrated on developing synthetic oil from "solvent refined coal," with loan and price guarantees for industry. Such actions could stimulate synfuel (oil and gas) proposals that have lain dormant in the West during recent years.

Even though the Carter plan has attempted to address difficult energy issues, an assessment of the full dimensions of federal energy policy, as it applies to the region, is almost impossible because there remains no comprehensive national energy policy. Project Independence, the Southwest energy plan, the North Central energy plan, the Prototype Oil Shale Leasing plan, and President Carter's National Energy Plan, and various proposals for nuclear production, solar prmduction, and geothermal and synthetic fuel production, each occur as main parts of federal energy policy. The Department of the Interior, the former Federal Energy Administration, the former Energy Research and Development Administration, and now the U.S. Department of Energy have continually revised their assumptions and projections. In a report in January 1975, Secretary of the Interior Morton revealed a previously undisclosed new scheme for power development in the region. An energy park, under consideration for eastern Montana, was proposed to be a 1,000-acre complex of gasifiers, converters, and refineries to produce large supplies of pipeline gas, synthetic oil, butane, and propane.[41] Morton brought both confusion and relief to Westerners when he announced scaled-down projections of coal supplies to be drawn from the region. Likewise, the Carter National Energy Plan has altered expectations of Western energy development. Carter's proposal has reduced some anxiety, but the major challenges to the West's economy and life-style persist.

Equally difficult to assess are the full environmental, economic, and social implications for the region. The original position of the Department of the Interior toward massive development in the region was expressed in its Final Environmental Impact Statement on developing federally owned coal in the Eastern Powder River Basin:

> Large scale stripmining in coal-rich Wyoming cannot be avoided under present plans and proposals, but mining and associated developments will have significant adverse environmental consequence. . . .[42]

Especially during the Nixon and Ford presidencies, environmental consequences were of little concern to federal and industrial developers in the region. The lack of ecological sensitivity in the North Central Power Study and Southwest Energy Complex plan and the noncompliance of federal agencies with the National Environmental Policy Act of 1969 suggest that development at any cost was the standard. The dedication to continuous economic growth, whatever the cost to the persons and regions upon which the physical impacts fell, was common to federal officials and large corporations.[43]

During 1977 and since that time, there have been indications of growing sensitivity toward environmental problems on the part of some key federal agencies. The energy Minerals Allocation Recommendation System (EMARS) of the Bureau of Land Management has been progressing slowly. Its duties now include monitoring site reclamation and rehabilitation activities. Passage of the Federal Coal Leasing Amendments Act over President Ford's veto in 1976 gives more discretion to states to expend royalty funds on the provision of front-end capital, site regulation expenditures, and environmental management. In addition, the recent Cooperative Agreements defined in Title 30, Section 211.75, of the Code of Federal Regulations enabling states to effect principle initiatives in regulating environmental standards on federal lands demonstrate a greater federal sensitivity to state demands for self-determination of their futures.

IMPACTS DUE TO DEVELOPMENT

As the primary interest here is in the environmental implications of widespread energy development, the full range of the impact on economic, demographic, and social elements is beyond the scope of this book. The multiplicity of development programs, their overlaps, and the uncertainties about their rates of implementation also hinder assessments of regional impacts. Implementation of these programs is furthermore subject to numerous potential inhibitors and constraints such as market fluctuations, changes in government policy, availability of capital, application of alternative energy technologies, role of national energy conservation efforts, political developments in the states of the region, and results of pending and perhaps future litigation on the part of environmental, agricultural, and Indian interests.

The environmental impacts of concern are all those emanating from the interrelationships between natural systems and technical systems in the region. Technical systems such as urban settlements, industrial complexes, communication and transportation networks, and extraction-conversion-transmission systems affect the natural biotic, hydrological, and atmospheric systems of the region. The use of land for surface mining, agriculture, water storage or diversion, in situ oil shale processing, industrial waste disposal, or human settlements leads to a range of environmental consequences. In the region the development of energy is inextricably related to water resources, land use, and population growth. A brief checklist of the environmental impacts of the region would thus include the changes in hydrology and geology, water supply and quality—surface and underground, atmospheric factors, natural systems of wildlife and open space, aspects of human health, and aesthetic values and cultural traits.

Within a few years of the commencement of energy developments expected to continue and expand enormously through the year 2000, there have been numerous indications of environmental degradation in the region. Experts trace most causes to energy-related industrialization. Emissions from coal-fired plants and the buildup of trace elements in surrounding areas have already led medical researchers to conclude that a great deal of the Southwest now registers significantly high levels of pollutant burdens in human tissues and increases in emphysema. The multifaceted symptoms of the boomtown syndrome first defined in Rock Springs, Wyoming, have been repeated in rapid-growth communities throughout the region. Landscape disfiguration and the destruction of agricultural lands have drawn protest in a number of states. Excessive industrial demands placed on the region's major river basins, plans for major diversion and storage projects, and the lowering of water quality by increased salinity, sedimentation, and acid mine drainage have caused alarm in communities, traditional Western industries, and government. The unlikelihood of successful revegetation of surface-mined land in the semiarid West is a critical environmental problem. The changes that demographic, industrial, and urban forces have begun to bring about in the rural-based economy and to the Western life-style have challenged many traditional social and cultural values.

A closer examination of the water, atmospheric, land use, and population impacts gives a more precise description of the region's environment and its susceptibility to change. Water-use patterns caused by large-scale development will likely be the most important determinant of environmental change in the region. There is general agreement among those familiar with the water resources of the region—and their scarcity—that there is not enough water to permit the congregation of coal-fired generation, gasification, liquefaction, and oil shale plants envisioned in existing development programs.[44] The agricultural economy of the region depends heavily on the water supply allocated to it and requires room for expansion in the future. Developing the oil shale industry to its potential three to five million barrels per day may require all the available water in the Colorado, Utah, and Wyoming area where the shale exists.[45] These general assessments of the water problem illustrate the environmental impact of vast water use and also indicate the implications for agriculture, future population growth, and the economy of the region.

The Colorado and Missouri river basins are the main components of the drainage system of the region. The surface water resources in basins are largely committed.[46] Diversions to energy industries mean denying supplies to other beneficial uses.

The uses estimated as necessary by the major development programs vary. The North Central Power complex was to require 2.6 million acre-feet per year, not including evaporative losses from hydro impoundments, requirements for support facilities, and urbanization. The Yellowstone River Basin, as the principal source, had an average of 10 million acre-feet flow per year, and 4.9 million acre feet were already committed to other uses. A diversion of 2.6 million acre-feet would reduce the flow of the river by 30 percent and in dry years by 50 percent. The estimate of the NGPRP further reduced water flow by the year 2000. With an estimated 12,800 megawatts of coal-fired electric generating capacity and another sixteen synthetic gas plants using water, total flow could be reduced to less than a million acre-feet per year. This estimate did not include supplies for revegetation programs and slurry pipelines. The NGPRP concluded that the Yellowstone Basin should be considered an area where available stored water is scarce, although the mainstream of the Missouri River (Ft. Peck, Sakakawea, Oahe reservoirs) had ample storage capacity of 66 million acre-feet. Deep groundwater supplies in the limestone aquifers known as the Madison Group are a potential source in the Northern Plains area of the region. For oil shale processing, retorting and refining an estimated 150 thousand acre-feet per year draw (for one million barrels per day production level) would place a serious drain on the Colorado and Green rivers.

The scarcity of water in the southern portion of the region is equally serious. The energy complex in the Southwest would require an estimated 454,000–594,000 thousand acre-feet per year by 1990. Nevada faces water deficits if population projections reach anticipated levels, Utah has very little water for future allocation needs once energy consumption estimates are totaled, and Arizona's Colorado River water is largely committed, as is New Mexico's. The estimates of water needed for the Navajo Indian Irrigation Project exceed the state's entitlement from the Colorado River.[47] The Bureau of Reclamation has considered weather modification, cloud seeding, and snowfall augmentation to boost water supplies in the Colorado River basin, although these activities may also mean adverse environmental effects such as flooding.

Another aspect of the water-related environmental problem is the changing of water quality. Degradation of water quality manifests itself in surface and underground water systems. Increased salinity, sedimentation, temperature, and the buildup of trace elements in water limit the range of uses for it and may adversely affect livestock and biotic and aquatic life. Wet cooling towers that recirculate water after use in power plants could return the water to rivers with significant temperature increases—a phenomenon that harms aquatic plants and fish.

The use of the surface and subterranean waters of the region leads to patterns in the establishment of industry and the location of human settlements, developments which have widespread environmental consequences. The way water is used in one state may also directly influence the water-use options of a neighboring state. Fouling the water in one area may lead to the degradation of underground supplies used in another part of the region. Environmental problems of water use extend beyond the problems of the site itself, and because of this, decision making for water use can rationally be made only within the context of regional development planning.

Consideration of the regional nature of environmental problems is also appropriate with regard to air quality. Research has shown that regions most susceptible to damaging climatic change are the steppe or semiarid zones such as the Rocky Mountain area. The Institute of Ecology has stated that "the magnitude of coal development proposed for the American West has the potential for degradation of air quality on a regional scale."[48] Dust from coal, oil shale, and uranium mines, and the emissions of air pollutants from electrical plants, gasification plants, and urbanization pose a serious threat to air quality in the region.

Two major consequences of regionwide energy developments which may lead to detrimental and irreversible effects on weather and climate are increases in the particulate loading of the air and large-scale changes in the properties of the natural land surface.[49] Increases in the pollutant levels in the North Central area, based on emissions from a concentration of the ten proposed 10,000-megawatt plants alone, are expected to be on the order of ten tons each, daily, of sulfur dioxide and nitrogen oxide. In addition, twelve hundred tons of particulates daily, twenty-four million tons of carbon dioxide daily, and huge tonnages of trace elements would pollute the region's air and could flow into Illinois and Minnesota. These trace elements, particularly mercury, lead, cadmium, flourine, boron, manganese, and beryllium, are already found in huge quantities in the air mass over the Southwest. The Southwest, traditionally known for its clear air, is today subject to emission loads comparable to the air pollution in Los Angeles and New York City.

Weather modification, such as variations in local climate, summer precipitation anomalies downwind from urban areas, and the altering of drought patterns are increasingly linked by scientists to human activities, particulate loading of the air, and a complex chain of reactions based on changing the surface of the land. The consequences of releasing sulfur dioxide—poisonous to human, plant, and animal life—occur in its reaction with water in the atmosphere, which can lead to sulfuric acid rains that damage vegetation and crops. Nitrogen oxide combines with ozone and carbon in the air to form smog, which reduces visibility. Even after pollution control devices such as electrostatic precipitators, scrubbers, and baghouse filters have removed 99 percent of these emissions, there is still a serious degrading of air quality in the region. The threat to human health is more serious than reduced visibility, crop damage, polluted surface and underground waters, and weather modification. Indeed, the Environmental Protection Agency has indicated that trace elements may "increase pollutant burdens in human tissues, alter human physiology, induce, aggravate or accelerate dangerous chronic disorders including heart disease, renal disease, lung disease, and cancer."[50]

Changes in the properties of the natural land surface were mentioned above as determinants of change in weather and climate. The way land is used is a fundamental determinant of environmental quality. Considerable acreages of the region will be changed as a result of surface mining for coal, perhaps for oil shale, and to a lesser extent for uranium. Estimates of surface mining for the 200,000 megawatt North Central Power complex totaled 50,000–175,000 square miles over thirty-five years.[51] Transmission line acreages were estimated at 4,800 square miles.[52] Transmission line mileages in the Southwest were estimated at 6,600 by 1985, but as widths of right-of-way are not established, conversion to the area to be occupied is not possible. The NGPRP acreage calculations for surface mines and some ancillary developments for CDP II by the year 2000 reach 82,518 acres. Eighty thousand tons of processed uranium ores will need to be dumped in the region by 1985, and perhaps 50 square miles of oil shale tailings will need relocation by the same year.[53] From a combination of measures from two studies, the total occupied surface acreage in the region for coal mines and ancillary plants between 1972 and 1985 could be almost 500,000 acres.[54] Towns and associated urban acreage increases are not included in this estimate.

In addition to these surface impacts, the NGPRP calculated losses in habitat used by major species of big game and prairie birds to total 430,000 acres by the year 2000 in the northern five states of the region. The one million acres to be disturbed for mines, associated facilities, and expanding or new communities is roughly half the nearly two million acres that was occupied by urban areas in the region in 1969[55]—an increase of 50 percent. In contrast, only 20,000 acres of the region were disturbed by surface mining between 1920 and 1970.[56] These estimates of surface disturbance are especially alarming in the context of findings of the National Academy of Science that indicate the rehabilitation potential of Western lands to be very low.

The environmental problems faced by the region are much larger than site-specific impacts on surface mining operations. The interconnections between water use and pollution, reclamation problems, and thermal pollution of waterways, on the one hand, and, on the other, heat from cities, respiratory and heart ailments, disposal of radioactive waste products from uranium mining, weather modification, and altering or fouling underground water systems have regionwide ramifications. Population increases as a result of coal, oil shale, and uranium developments could reach 600,000 people by 1985, a 5–6 percent increase of the region's population over ten years. The economic growth and development generated by this increase, however, and the environmental problems they bring have already caused much alarm. Equally alarming is the random and uncoordinated way in which development has proceeded.

ENERGY PROBLEMS BY STATE: VIEWS OF STATE OFFICIALS

This section of the chapter will deal with the differences in state postures on energy problems, differences that any regional policy mechanism must confront in its efforts to articulate policy for its constituents.

As indicated in Chapter 1, it was through the formation of the Western Governors' Regional Energy Policy Office (WGREPO) that the states displayed their major political, philosophical, and policy concerns. Each state's representative to WGREPO—it was usually the chief energy spokesman for the state—was asked three questions by the author in researching this book.

(1) What are the principal energy-related problems in your state?

(2) Which energy-related problems may be more effectively addressed by cooperation with neighboring states?

(3) What does your state see as the utility of a ten-state coalition on energy policy?

The questions were designed to shed light upon similarities and differences in principal problem areas as well as to identify some "scope" in the problems of each state. Some energy-related problems can be addressed within state boundaries, but there are others that transcend political borders. Accordingly, energy officials were asked to define these and to indicate the primary purposes of a multistate energy organization. Reponses were not prioritized with the exception of North Dakota's and South Dakota's on the first question.

From the responses—detailed in the summaries at the end of this chapter—several conclusions can be drawn. The first question generated the greatest disparity of responses, while the last question saw the most similarities. The Northern Plains states of Montana, North Dakota, South Dakota, Wyoming, and Colorado were more sensitive to the problems referred to in the second question.

Some of the states are clearly more conscious of environmental quality issues than others. Montana, Colorado, and Wyoming each stressed the need for environmental preservation, whereas Utah, Arizona in particular, and South Dakota to a lesser degree

TABLE 1
Expectations of Energy Facility Expansion:
Preliminary, planned, or proposed new or expanding energy facilities in Federal Region VIII*

	Coal Mines		*Electric Power Plants*			*Coal Conver-sion*	*Bio Conver-sion*	*Oil Shale*
	U.G.	*Strip*	*C.F.*	*Nuc.*	*Other*			
Colorado	46	31	6	0	4	1	0	7
Montana	0	12	4	0	5	5	0	0
North Dakota	0	11	3	0	0	3	0	0
South Dakota	0	0	5	0	4	1	0	0
Utah	31	3	6	0	4	1	1	3
Wyoming	6	31	5	0	1	5	0	0
Regional Total	83	88	29	0	18	16	1	10

*Source: Department of Energy, Department of Interior, Bureau of Mines, *Projects to Expand Energy Sources in Western States* by Charles Rich, 1977

described this area of public affairs as not of major concern. In North Dakota and South Dakota the preservation of land and the environment was interpreted more in the context of the protection of agriculture.

The opinions expressed regarding the utility of interstate cooperation and negotiations also registered great differences in view. One of South Dakota's principal motivations for interstate negotiation with contiguous neighbors was to ensure energy supplies. Similar reasoning was expressed on behalf of Nevada and Arizona. Colorado, Nebraska and South Dakota each expressed fear of impacts from the development projects of neighbors crossing borders and desired channels of communication with immediate neighbors. In contrast, Montana and North Dakota in particular sought the benefits of sharing information, experiences, and research findings with contiguous neighbors.

From the answers to the third question it is clear that the majority of the states agree in principle on the wisdom of a ten-state coalition. There is general agreement that the West as a region, and each state individually, may benefit from a unified approach to some energy-related problems. The principal objective of unification on policy matters is to seek power.

Tar Sands	*Geo-thermal*	*Coal Slurry*	*Uranium Mines*	*Oil Refin-eries*	*Uranium Mill & Enrich-ment*	*Natural Gas Proc. & Storage*	*Rail-roads*	*Pipe-line Petr. N.G.*
0	0	1	30	3	8	0	2	2
0	0	1	0	0	0	0	0	0
0	0	0	0	0	0	0	Nu-mer-ous	1
0	1	0	1	0	1	1	0	1
2	2	1	21	0	0	3	1	1
0	0	2	15	3	4	1	1	0
2	3	5	67	6	13	5	4+	5

The states wish to increase their leverage in decision making for energy at the national level. The most obvious areas of concern in substantive terms are to seek development impact assistance funding, to entice energy research and development funds into the West, and to defend the rights of states in matters of water use, energy facility siting, and the management of environmental regulations in mining areas.

The views expressed by senior energy officials in state government have remained relatively constant since 1974, even though personnel may have changed. Today the Department of Energy continues to project broad intentions for major mining and industrial development in the region. As table 1 indicates, Project Independence has been scaled down, but the Carter Energy Plan continues to project severe impacts for the Rocky Mountain states. The prospect of eighty-eight strip mines, twenty-nine coal-fired generating plants, sixty-seven uranium mines, and, in Colorado alone, an assortment of dozens of other power plants has generated deep concern there.[57]

Jim Monaghan of Colorado, one of the principal actors in the evolution of Western mechanisms for energy policy, is convinced that federal attitudes regarding development have not changed significantly since 1974 during the course of his incumbency as a senior adviser on energy and natural resources to the governor of Colorado. States Monaghan,

> There is no real guideline or roadmap for the West to follow. There is no more clarity in federal plans with Carter than there was with Nixon and Ford. The game hasn't changed much. The mix of policies is relatively constant, as the same technicians are working on them in Washington. The *acceleration* [emphasis added] in development out West is still expected . . .[58]

Both Governor Richard Lamm of Colorado and Governor Thomas Judge of Montana were still as concerned about the economic, environmental, and sociological health of the region in 1978 as they were in 1974 and before.[59] The impact of energy development and population growth poses both enormous challenges and opportunities for the West.

The responses of the state WGREPO and WESTPO representatives to my three questions follow.

Alaska

Principal energy-related problems

1. Energy costs are astronomically high, varying markedly from region to region, even though the state is a storehouse of resources.
2. The state aims to develop the energy resources without adversely affecting social and economic and environmental conditions. Reserves of oil, coal, and hydropower are all in isolated and undeveloped areas. Provision of capital for communities, schools, transportation, and the like is very costly and not supplied through national energy policies.
3. The bulk of Alaska is public land. Proposals of the U.S. Department of the Interior for distributing lands and recommending land use are in significant conflict. The state has been preoccupied with the problems and conflict surrounding these issues for more than two years, almost to the exclusion of other major public problems.

Energy-related problems more effectively addressed by cooperation with neighboring states

1. Alaska is not contiguous with any state but sees advantages in collective behavior with some Rocky Mountain states who face similar problems.

The utility of a ten-state coalition on energy policy

1. Collective action on behalf of sparsely populated Western States is potentially advantageous for Alaska because of its small congressional delegation. Alaska, the largest state in the union, has little political clout in Washington, D.C.
2. Alaska finds a number of Rocky Mountain states, especially Colorado, confronting similar energy-related problems. Communication is advantageous in this situation.
3. Alaska's difficulties in influencing public lands decisions through the Department of the Interior could be a test case for Rocky Mountain states in the future. Alaskan officials believe they can aid the Rockies and in turn benefit from Western states' support in natural resources policy making.

Arizona

Principal energy-related problems

1. Because Arizona is a consumer state—imports approach 85 percent of consumption—the principal problems are guaranteeing supplies, needing more exploration, and developing alternative sources such as solar power.

Energy-related problems more effectively addressed by cooperation with neighboring states

1. New Mexico has instituted an energy tax that discriminates against some Arizona consumers who import energy from across the border.
2. The state exports electricity to California and imports it from Colorado, Utah, New Mexico, and Texas. Such a network of generation flows could benefit from interstate cooperation.

3. Water scarcity is a serious problem and will require interstate negotiations.

The utility of a ten-state coalition on energy policy.

1. The primary interest to Arizona is to further communication with the federal government. There is more need for this than for cooperation between the state and Northern Plains states.
2. The interstate mechanism may help resolve the competitive stands currently adopted between most of the states and to lure federal energy research and development funds within reach.

Colorado

Principal energy-related problems

1. The state's power to control the rate and amount of energy development within its borders is ineffective.
2. The state needs a stronger voice in the public and private arenas that control energy policy decisions.
3. The cumulative socioeconomic impacts of oil shale development proposals could be very serious.

Energy-related problems more effectively addressed by cooperation with neighboring states.

1. Water scarcities must be addressed on a multistate basis.
2. Oil shale development in Utah affects water quantity and quality in Colorado and creates migration of people that will strain each of the two states.

The utility of a ten-state coalition on energy policy.

1. Many of the tradeoffs required to accommodate both energy development and preservation of the quality of life will have to be made on the regional level. These decisions require multistate cooperation.

2. Responsibility for energy production should be divided up equitably between the states in the West. Such decisions require multistate cooperation.
3. The states in the West need to address federal resource policies with a unified voice because of the advantages of political power.

Hawaii

Principal energy-related problems

1. Hawaii does not share the problems of energy development, water policy, and agriculture that face the other Western states.

Energy-related problems more effectively addressed by cooperation with neighboring states

1. Hawaii has no neighboring state, but does see itself as a Western state. It confronts the common problems of political disunity in national policy making posed by sparse population in an area of rapid economic expansion.

The utility of a ten-state coalition on energy policy

1. Where public issues, whether related to energy or to any other subject, become set in a context of West vs. East, Hawaii wishes to support other Western states and to receive political support in return.
2. Hawaii is dissatisfied with the degree of flexibility in the implementation of many federal programs and seeks multistate collaboration in efforts to pursue greater state autonomy in decision making.
3. Hawaii seeks multistate support in advancing the prosperity of the Pacific Rim states and enhancing the image of the West as the major link between the United States and the Asia-Pacific region.

Montana

Principal energy-related problems

1. Federal executive agencies increasingly act to preempt states' rights on energy-related matters, especially the power over water use and energy facility siting.
2. Major difficulties are associated with trying to retain control over the state's resource development and to retain decision-making authority.
3. Stress falls on the state's legislative and executive functions when they must analyze, process, and regulate mounting programs for coal development.
4. Planning orderly growth in the state is difficult in the absence of national energy and growth policies.

Energy-related problems more effectively addressed by cooperation with neighboring states

1. Impacts related to air quality, agriculture, environmental quality, and social and economic stress all cross state borders.
2. Water scarcity and sharing the water of the major river system of the Northern Plains subregion is another such problem.

The utility of a ten-state coalition on energy policy

1. The threat of preemption of states' rights, and the lack of state input to federal decision making, require the unity of the region's states with a view to gaining leverage for more consideration in federal policy making.

Nevada

Principal energy-related problems

1. As a non-fuel-producing state, our principal concern is to ensure supplies.

2. The state objectives are to encourage orderly growth, to diversify the economy (from heavy reliance on gambling), and to ensure that energy is available to carry out these objectives.
3. Strip mining, pollution, and energy-related impact problems found in other states of the region are not major Nevada problems.

Energy-related problems more effectively addressed by cooperation with neighboring states

1. This state has not experienced any at this point.
2. Potential problems of multistate air quality management and the arbitration of discriminatory interstate energy tax programs may require negotiations with neighbors in the near future.

The utility of a ten-state coalition on energy policy

1. States undergoing similar energy problems and seeking administrative solutions are provided with information.
2. The mechanism may facilitate negotiations between states as to who will provide which of the various energy types and which states can do so most efficiently and quickly.
3. A unified Western bloc has more potential clout in seeking national responses to its needs.

New Mexico

Principal energy-related problems

1. Water scarcity for energy development, Indian development projects, and expansion of the state's economy generally are energy related.
2. Financial capability of the state to develop energy sources is very limited.
3. The social and environmental impacts of large-scale energy projects are usually a problem.

Energy-related problems more effectively addressed by cooperation with neighboring states

1. Water scarcities are crucial. No energy development proposals can be made by any one state in the Four Corners area without planning in concert with the other states and federal agencies.
2. Air pollution problems may grow more widespread and serious in the Four Corners area.

The utility of a ten-state coalition on energy policy

1. The states will be enabled to deal more effectively with federal agencies and to bring a Western influence into national policy setting.
2. Information and energy-related experiences can be shared with other states in the West.

North Dakota

Principal energy-related problems

1. Mined land reclamation must be controlled and supervised to completion.
2. State rights over water use must be maintained.
3. North Dakota must cope with the management and share responsibilities for federal lands and federal minerals in the state.
4. Social and economic problems are thrust on small rural communities by massive energy developments.
5. Maintenance of state air quality is a problem.

Energy-related problems more effectively addressed by cooperation with neighboring states

1. Problems of water allocation and use require cooperative solutions.

2. Each of the neighboring states has to deal with federal land and federal mineral development within its borders. Sharing data and experiences is helpful.

The utility of a ten-state coalition on energy policy

1. Sharing information on common problems is helpful.
2. As each of the region's states faces serious water scarcities and has a large portion of state territory under federal public land control, the common difficulties may be eased by cooperation.
3. National decision makers can be pressured, thus enhancing the probability of the West receiving consideration in crucial energy policy decisions.
4. A united multistate front can approach federal sources for impact funds related to energy development.

South Dakota

Principal energy-related problems

1. Adequate fuel for home heating and essential facilities must be provided.
2. Adequate fuel for the state's agricultural industry is essential.
3. The secondary impacts (water and air pollution, diminished water quality, damage to critical environmental areas such as the Black Hills, social and economic impacts stemming from boom communities in neighboring states) caused by energy developments in Wyoming, Montana, and North Dakota affect South Dakota.

Energy-related problems more effectively addressed by cooperation with neighboring states

1. Each of the problem areas above could best be so addressed.

2. The state is an importer of energy, and neighboring states are exporters. The states need their fuels, but the unfavorable impacts induced by their developments are a problem. The state cannot control the pollution of neighbors nor guarantee supplies.
3. Some South Dakotan suppliers of power are constructing plants in Wyoming.

The utility of a ten-state coalition on energy policy

1. To the extent that the West can be an effective voice in the setting of national energy policy and related matters, that is justification for regional action.
2. There are considerable benefits from sharing the experiences of neighbors in energy policy matters. The shared experiences and information transfer help spread innovation in problem solving.

Utah

Principal energy-related problems

1. The state is eager to develop its energy resources, particularly oil shale, in an orderly manner. The problems of socioeconomic impact in boom communities are of concern.
2. Many of the energy resources of the state are on federally controlled public lands, and the state has experienced difficulties in attempts to be a partner in decisions regarding development proposals emanating from federal authorities.

Energy-related problems more effectively addressed by cooperation with neighboring states

1. The state sees little problem in air pollution, water pollution or scarcity, or landscape degradation.

The utility of a ten-state coalition on energy policy

1. The principal use of united action on behalf of Western states is to demonstrate what the problems and needs of the region are.
2. Sound baseline data on energy matters are urgently required in the West, and new regional organizations may assist in generating them for the states.

Wyoming

Principal energy-related problems

1. The trade-off between energy development and the maintenance of environmental quality is a problem. The administration of Governor Herschler is strongly committed to environmental preservation.
2. Social and economic problems are caused by unplanned and rampant development in some communities that are not equipped to handle the growth effectively.
3. Internal supplies of natural gas are restricted and fall short of demand. The state exports 78 percent of its natural gas.

Energy-related problems more effectively addressed by cooperation with neighboring states

1. Air pollution and thermal effects of large generators may affect precipitation in states downwind such as South Dakota and Nebraska. If so, Wyoming will need to alleviate the problem by appropriate attention to state air quality standards.
2. There is movement of population between domestic and employment locations that are separated by the Montana-Wyoming border. Taxing arrangements may need to be negotiated between the states to satisfy community fiscal demands.
3. Water sharing negotiations are needed between states.

The utility of a ten-state coalition on energy policy

1. A coalition would facilitate the exchange of ideas and experiences of states facing similar energy-related problems.
2. A united Western front is needed to apply pressure for the establishment of sound national energy and resource policies.
3. There is a need to encourage the appropriation of research and development funds for use in the region that is expected to carry a major burden of energy production for the nation.
4. Innovative institutional arrangements are needed in the areas of state-state relations and state-federal relations.

Toward Understanding the American West and Its Experiences as a Region

3

This chapter provides a brief background for understanding the West's history, life-style, unique characteristics, experiences with the federal government, and experiences with regional organizations. The chapter traces and reports on a selection of events in the West during the 1970s that indicate a gradually increasing awareness on the part of public officials of the severe implications for the region should energy development proceed as proposed. In addition to concern for the environmental, economic, and social impacts which would be brought about by population growth and industrialization, it became clear that certain problems would be unmanageable without a multistate approach.

There are four parts to this chapter. The first reports on the West, its land, the Western life-style, and environmental quality. The distinctive characteristics of this vast area of low population density, immense size, and little industrial activity are examined, and the findings of scholars who have analyzed the regional and sectional attributes of the West are reviewed. Included also is an examination of the thesis that the West has long been the object of discrimination by the federal government.

The second part of the chapter describes the West's experience with regional organizations. This brief discussion will consider the types of organizations in the region and their range of policy matters.

The third section focuses on states' responses to the issues of rapid population growth, economic stimulation, and industrialization from energy development in the 1970s. These responses, in a number of instances, point to a type of public and official acceptance different from that traditionally expressed in the West prior to 1970.

The final section of the chapter examines the basis for increased use of multistate mechanisms, with particular emphasis on the Northern Plains states. The section briefly identifies a representative selection of problems Western states have faced since 1970, problems which were predominantly instigated by federal action and which the states have had little opportunity to address in the context of federal decision making.

THE WEST: LAND, LIFE-STYLE, ECONOMY, AND REGION

The reaction of the state of Wyoming to the threat of massive energy development was the subject of a recent study by John Jenkins, a political activist in the state and an astute observer of rapid change in Wyoming and contiguous states. In his analysis, Jenkins wrote of the distinctive features of the American West:

> To understand life in the little-developed West, and why its inhabitants might want to preserve it, we must develop a sense for the land, the economic activities of its inhabitants, the nature of day-to-day life, and the hopes and goals those people hold dear.[1]

The land is vast, with a multiplicity of land forms. The diversity of climates in the region, extremes in rainfall, dryness, and temperature are not matched anywhere else in the United States. The West's richness in natural resources, such as forests and minerals, combined with its variety of crops and large agricultural production, mark it as a distinctive part of the continent.

This great section of the United States is relatively untouched. There is little industrialization, and the West's population density of 9.7 persons per square mile is in stark contrast to the 953 in New Jersey. The Western economy, however, does not exact heavy penalties for the advantages of low industrialization and low urbanization. As Jenkins puts it, photosynthesis may be the most important economic activity of the region.[2] The life of most Westerners has traditionally been dominated by size, open space and natural areas, neighborly rural communities, and informal grass-roots politics and government. Jenkins refers to the West's culturally and geographically distinct segment of the American public; Professor Frank Jonas terms the West "a region with an underlying unity in physiographic features, economic impulses and political objectives."[3]

Although few would argue with these descriptions, whether the West has ever exhibited the attributes of regionalism or sectionalism is subject to debate. Different perspectives on the West were developed in the theses of Walter Prescott Webb,[4] Carl Frederick Kraenzel,[5] and Frederick Jackson Turner.[6] Turner emphasized the significance of the frontier and the role played by geographic and cultural sections in American history. Webb's notion of geographical determinism and Kraenzel's thoughts on migratory frontiers, geography, and institutional arrangements add thoughtful dimensions to an understanding of the West. It is valid to claim the West as a geographical reality, and a culturally distinct region as well, but there is no general agreement about the many other aspects of supposed regionalism. John Walton Caughey, noted Western historian, provided a historical interpretation of the West that is a plausible place to begin.

There is little doubt that the West today is in large part the product of its history. According to Caughey, the West has had a sharply stratified character—a term which implies that many of the Western states have passed through similar historical periods:

> . . . the Indian period, followed by an epoch of European contact and limited occupation; the early nineteenth cen-

> tury invasion by American explorers, traders, missionaries, and empire builders; another era dominated by miners, farmers, cattlemen, Indian fighters, and railroad builders; then a generation dedicated to making the most of the railroad, but with attention also to the tourist, the health seeker, real estate speculation, and developing the eastern markets; and next our own more sophisticated generation, with a more far-reaching application of science to the problems of production, processing and transportation. . . .[7]

To be added to this incrementally evolving history is the present period of major development projects for irrigation, hydroelectric power, water diversion, and energy exploitation. Most of the states in the West have experienced these activities directly, some more indirectly.

Caughey's claim is that an unmistakable regional unity shows through this history. This unity may be better expressed as a commonalty of experiences, particularly with regard to economic and governmental affairs. As Caughey correctly indicated, "It is hard to see how anyone can take stock of the American West since 1848 and not see growth and development as the dominant characteristics."[8] The economy has been primarily extractive. It began with furs, then went to sandalwood, cowhides, gold and silver, on through great varieties of items to, more recently, oil and gas, and now uranium, coal, oil shale, and geothermal energy. There can be no argument with the statement that the West's role in the national economy has been as a major supplier of raw materials. In Caughey's terms, the West has been a kind of colony. In recent years views similar to Caughey's have been frequently espoused by scholars.

While the economy produced bountifully and while in general the region welcomed growth and economic benefits, state governments played inconspicuous roles. With nearly 56 percent of the land area of the West in federal ownership and the federal government as the only source of the vast investment capital needed for large development undertakings, the influence of the states of the region ebbed and flowed with federal action and inaction. In fact, the most prominent strains in the history of the American West are those of federal economic activity in the area.

> Not only has the economy of the West been persistently colonial, the plain truth is that the West is also a minority section. It always has been and it probably always will be. As such it naturally and inevitably has been discriminated against. . . .[9]

This discrimination is demonstrable in political and economic facts. The setting of state boundaries, some unreasonable delays in admission to statehood, and poorly timed political reapportionments in relation to population growth in the West are examples of political discrimination in the federal government. In the economic sphere, hidden taxes and indirect charges, inequitable freight schedules, and tariff laws have also discriminated against the West.

Western reaction to these circumstances has been scant, according to Caughey, with little general awareness of them.

> Curiously Westerners have shown relatively little awareness that they are cast in this subordinate role. In part, no doubt, the colonialism is not felt because it is concealed. . . . Western climate and scenery doubtless distract from the hard realities. . . . another good reason is that comparatively few Westerners have much realization that the West is a region. . . . the West has never been organized as a political unit, much less as an independent confederacy. . . .[10]

The political barriers to effective protest on behalf of the West—not that prior to 1975 there was significant protest to federal authorities or the nation—are real. The power center of the nation is far removed in the East, and, as the region is defined here, the West constitutes less than 5 percent of the American vote. Representation of the states in Congress is even smaller. According to the economic theory of competition, the West has been a price-taker, not a price-maker. In other words, it has more often than not danced to whatever tune the political, governmental, and financial powers played.

Given the commonalty of life-style, economy, and history in the West, one asks whether the West has exhibited any characteristics of regionalism or sectionalism. The West was once the most distinctive of American sections, according to Professor Edgar Lane, long-time student of Western politics.[11] He contends that the West has had its sectional values eroded over time, as have all the nation's once sectionally minded areas. The nation's domestic politics were once the "relatively simple domestic politics of sectional economic adjustment," but the "light of public life has drifted from the periphery [states] to the center."[12] Urbanization, modernization, and great population growth in the West he defined as the primary homogenizing factors. Nonetheless, he found that there existed a peculiar internal consistency in the West's voting patterns in both rural and urban areas, a casual slackness of its party organization, both a tradition of deference to the shibboleths of nonpartisanship, and yet periodic evidence of recurring party cohesion. To these regional peculiarities of politics, Professor Sharkansky, a prolific scholar of American politics, added, a decade later, high voter turnout and the typically unsophisticated, part-time state legislatures.[13]

Politics in the West today is marked by the qualities Lane and Sharkansky noted. According to Lane, "For Western Congressmen, party identifications appear increasingly to outweigh consideration of sectional interest."[14] Sharkansky, in agreement with Lane, addressed the strong historical combat between the two parties in the West and concluded that "no sectional trauma has aligned the residents of Mountain states into one party or another."[15]

From an analysis of Western political concerns and voting patterns in Congress, Western political observer and teacher Neal Maxwell concluded, similarly to Lane and Sharkansky, that "what we see . . . is clearly neither a requiem for a region nor a resurgent regionalism."[16] Maxwell probably would not have objected in 1970 to the claim that Lane made more than a decade before:

> The "political West" may be little more than a semantic convenience. Western issues are being crowded out of our field of vision. Western voters conform to national configurations. Western Congressmen cast party votes on issues that have lost their sectional simplicity.[17]

K. Ross Toole, well-known Western historian, has pointed out the absence of regional political solidarity in Western states, J. L. Plummer has recognized the challenge of maintaining Western political unity, and L. K. Caldwell and others have discussed the nature of colonialism and the West.[18] Jonas, in tending to contradict himself, brings final evidence of the traditional political disunity of the West. Stating at first that the Western states can be viewed as a region with shared political objectives,[19] Jonas concluded that "never have the 13 western states, nor even the 11 or 12 [with Alaska] continental states, voted in concert on a bill in Congress directly affecting the economic welfare of the area."[20] Only in the sense that these traditionally growth-oriented states have uniformly appealed for federal assistance to develop their economies have their political objectives been the same.

Lane concluded that there is no reason to believe that the political West, such as it was, will be reborn.[21] A survey taken in 1969 by Maxwell, however, gives some indication of the common interests shared by Western political elites. Maxwell's survey revealed the issues most important to Westerners in Congress, the most prominent political-economic issues requiring attention, and a list of issues around which Western politicians could possibly rally.[22] Problems of land and water development are the common issues of concern, in particular the federal land management policies pertaining to mining, reclamation, recreation, and grazing. One of the most persistent matters the West needed to address, according to the Western solons who were surveyed, was the obtaining of federal or private capital sufficient to develop the West's resources. These points of view are compatible with the traditional experiences of the West in its overriding endeavor to achieve economic stimulation. The chief regional rallying points were reclamation, resource development, mining, conservation, and recreation and tourism. The politicians who were polled saw the westward tilt of the nation's population as a serious problem, and yet they were convinced the future of the West would mirror the past in its search for development and growth.

Many common experiences and growth-oriented aspirations of Western states were noticeable in 1969 when Maxwell did his survey. Although some observers have encouraged a regional approach that would give the West more power relative to the rest of the nation, there has been little use of organized multistate approaches to problem solving. There has been a proliferation of multistate organizations, but few have pursued their goals in a "regional" fervor. The West could gain politically and economically from a regional approach to its problems, Caughey was convinced:

> regionalism need not be antisocial. . . . to achieve it we need to rise above provincialism, cultivate an awareness of the essential unity of the West, and strive for at least a measure of solidarity.[23]

THE WESTERN EXPERIENCE WITH REGIONAL ORGANIZATIONS

Organized multistate approaches to problem solving in the West may have been few, but within recent years numerous multistate organizations (MSOs) have sprung up. Prior to 1975 (the year of the formation of WGREPO), no one regional energy organization had as its focus the same boundaries as the ten-state region under consideration here, while several organizations included different sets of these ten states and their neighbors. A survey of these MSOs conducted by a Western Governors' Task Force in late 1976 discovered over a hundred organizations; subsequent analysis has shown, however, that formation of these MSOs was not matched with any sense of regional planning or multistate organizing.[24]

Seven kinds of MSO were defined and examined in the study. Sixty of the hundred MSOs consisted of interstate compacts, state executive branch associations, and joint federal-state regional organizations. Twelve MSOs dealt with energy, twenty-three with water issues, twenty with the environment, eighteen with education, and at least sixteen with agriculture. Only three MSOs were identified as having general crosscutting functional responsibilities. The narrow functional concerns of the MSOs, the excessive duplication of effort, and the near-absence of accountability to public officials reflected the fragmentation of Western multistate problem solving.

The organizations for economic development, river basin commissions, and certain interstate compacts shared by this group of Western states are worthy of further examination. The West has two federal-state regional organizations for economic development in operation. The older is the Four Corners Regional Commission (Four Corners), created in 1966, which encompasses a ninety-two-county area of portions of the states of Arizona, Colorado, New Mexico, Utah, and Nevada. The other is the Old West Regional Commission (Old West); it was established in 1972 and encompasses the entire states of Montana, Wyoming, North Dakota, South Dakota, and Nebraska. Both of these organizations were created pursuant to Title V of the Public Works and Economic Development Act of 1965.

The commissions are federal-multistate mechanisms. They are financed by the U.S. Department of Commerce together with relatively small state matching funds. Their purpose is to assist development activities so as to achieve a closer interregional economic parity. Programs include the drafting of long-range regional economic plans and the coordination of these plans through specific action programs. An average of $4.5 million has been spent for these programs by each commission each year.

A federal cochairman, with veto power, appointed by the president of the United States, and the governors of the participating states constitute each commission. Neither Old West nor Four Corners has authority over federal agencies which would enable it to plan regional development. Because of the limited funds they pass on to states, these organizations have found it difficult to claim the interests of governors. A survey of state officials participating in the nation's commissions revealed how limited in value the commissions were.[25] In regional planning, the commissions have little influence. They are of value to the states only insofar as they are conduits for federal funding of state and local projects. Governors grow concerned when commission funds have not been divided equitably between states.[26] In other instances, neighboring states have joined forces to study the need for regional facilities such as an airport.[27] At this time, commissions have not been operating long enough to assess for program effectiveness.

The Missouri River Basin Commission, which involves six of the states in the energy policy region, is another example of joint federal-state regional organizations. Formed in 1972, this commission includes Colorado, Montana, Wyoming, Nebraska, North Dakota, and South Dakota in addition to four other Plains states. The commission was created through the Water Resources Planning Act of 1961, Title II, and is a forum for federal executive departments to meet with state representatives to coordinate planning for river basin development.

This commission is a federal-state cooperation mechanism, and like the Old West and Four Corners models, was created through federal initiatives. It is not an independent decision maker for the region and has no authority over states or federal agencies. As a coordinator of agency planning rather than an independent planner and as a minor conduit of money from the federal Treasury, the commission has generated little interest from states.[28] Martha Derthick's analysis of certain forms of regional organization has concluded that states agree to participate in organizations like this one primarily for defensive reasons. Where interstate conflicts over water allocation are intense, as they have traditionally been in the West, states find it advantageous to defend their interests in such a forum; they at least can keep abreast of current information. There is less gubernatorial participation in this commission than in Title V, probably because of its lower potential for delivery of federal project funds. It also is not a general purpose organization as are Four Corners and Old West.

Among the MSOs that are pertinent to the concerns here, there are four categories. First is the interstate compact variety, and second, some of the mechanisms sponsored by the Council of State Governments. Third is the nonprofit corporation represented in the West by the former Federation of Rocky Mountain States; fourth, multistate political mechanisms.

Mechanisms for interstate cooperation are provided for in the Constitution of the United States. Formal agreements between state governments are enacted as statutory law and generally require congressional consent depending on the substance of the compact and the intent of the states. As formal arrangements, interstate compacts may require commitment of state funds as well as authority and action by each member state's legislature. Traditional functions of compacts involve settling boundary disputes or allocating water from river systems. Since World War II, compacts have been increasingly instituted for the sharing of facilities and finances in functional areas of public affairs such as higher education, corrections, nuclear energy, health services, and welfare.

Western states in this study are engaged in a multitude of interstate compacts: Montana, fifteen; Wyoming, twenty; North Dakota, sixteen; South Dakota, fifteen; Nebraska, sixteen; Utah, fifteen; Colorado, twenty-four; Arizona, fifteen; and New Mexico, twenty-two. These compacts sometimes link states within the region with states outside the region.[29]

The Colorado River Compact apportions the waters of the upper and lower basins of the Colorado River. Other compacts, such as the Western Interstate Corrections Compact, address the provision of public services, and there are those, such as the Western Interstate Nuclear Compact, that act as clearinghouses for information.

Within the vast array of regional instruments there is a subset affiliated with the Council of State Governments. These instruments provide staff support to governors, legislators, attorneys general, and state budget and planning executives for events like regional conferences. At such meetings regional policy statements are discussed. Research projects may be created and funded at the request of governors, an example being the Western States' Water Council. This research and information-generating instrument was created in 1965 at the direction of the Western Governors' Conference.[30] Such organizations are usually comprised of gubernatorially appointed state delegates. Participation is voluntary and few policy positions may be taken without the unanimous sanction of governors. The Western Governors' Conference itself meets annually to foster an exchange of views, and in particular to discuss regional development, and interlevel government communication. These groups serve primarily as mechanisms for information exchange, however, not really as formal organizations.

One state-initiated organization—complete with permanent structure and staff—was the Federation of Rocky Mountain States. The Federation was chartered in 1966 with the governors of Colorado, New Mexico, Idaho, Utah, Montana, and Wyoming as the incorporators and board of directors. Minor funding for the organization was contributed by each of the states, with the greater portion of funding coming from leading businesses in the region and the federal government. Originally the Federation was established for the promotion of economic development in the relatively underdeveloped Rocky Mountain states. More recently it worked at identifying multistate issues, attempted to develop regional policies, and implemented programs of public affairs such as land use, energy, transportation, and vocational education. As a link between governors and private industry, the Federation was largely an entrepreneur of economic stimulation in the six-state area. Even though a large proportion of its sponsorship lay with industry, by far the larger portion of its funding came from federal government grants.

In addition to gubernatorial and administrative involvement in regional organizations, there has been the activity of legislators and congressmen. Two multistate arrangements have been based on partisanship—the Conference of Western Democrats (1959) and the Western Regional Republican Conference (1965). Nonpartisan arrangements such as the Conference of Western Senators (1949) and the more recently formed Western Conference (1975) were also organized. According to Maxwell, the first three are not regularly active.[31] The Western Conference, the last of these to be created, is an informal and voluntary arrangement initiated largely at the hands of former senator Frank E. Moss (Democrat of Utah) and Representative Morris Udall (Democrat of Arizona). The organization has convened a number of times since 1975; it has the authority to draft policy resolutions regarding common concerns of energy development but to date has contributed little.

This brief examination of existing and past regional instruments in the West reveals six important findings. First, most of the vast number of compacts formed in the decade address the provision of various public services. Second, the policy issue of water planning and allocation has been of most concern. Third, the majority of interstate arrangements have advisory and service provision and planning functions, without authority to regulate member states' activities. Fourth, interest in multistate arrangements has been greater among governors than among administrators or legislators. Fifth, there has been a proliferation of various types of MSOs, especially in the 1970s. Finally, Western MSOs are fragmented, have duplicate functions, and are not organizationally conducive to regional problem solving. Evidence of a regional consciousness on the part of the initiators of these arrangements is scant; the intention is instead to secure profit for individual states. In accordance with the West's thirst for economic development, governors have shown most interest in the organizations that best facilitate the profit objective: the Federation, Old West, and Four Corners. With the nation's most massive industrialization programs directed toward these states after 1970, one would expect that the West would welcome the economic stimulation. The next section of this chapter examines the validity of such an expectation.

THE WEST OF THE 1970S

There is no doubt that the ten states covered in this study have historically been motivated by the desire for economic development and in a modified way will continue to be. Around 1970 qualifications were attached to the pursuit of growth in some of these states—in some more than others. Similarly, in the early 1970s, the awareness of environmental consequences of population growth and industrialization increased noticeably in some states. The changing attitudes have been reflected in the changing positions of state governments and in growth-related laws passed in state legislatures. There are at least four prominent instances in which states have attempted to direct their futures by controlling growth: Colorado's changing attitude toward environmental preservation, Montana's rapidly enacted energy-related legislation, Wyoming's rapid adoption of land use planning principles, and North Dakota's regional environmental assessment program each attest to attitudinal change. In contrast have been the postures that Arizona, New Mexico, and Utah have adopted toward growth control, environmental protection, and energy development.

As early as 1966, Colorado was making a serious bid for the 1976 Winter Olympic Games. At this time the state was still clearly in a progrowth phase in its history. The Republican following in Colorado was strong, prodevelopment sentiment was pervasive, and Governor John Love, elected in 1963, was reelected twice on a "bring-business-to-Colorado" platform. But in the late 1960s, while ski lodges such as Vail, Aspen, and Steamboat Springs boomed and while national firms were being lured to Colorado, antigrowth campaigns and surges of an environmental consciousness appeared for the first time. While there was protest over the routing of Interstate Highway 70 through Eagle's Nest Wilderness, the Colorado legislators were lobbying in Washington, D.C., for a number of reservoir and water diversion projects.

In 1971 the indicators of change in the public attitude toward ecological matters were evident, and when the election of November 1972 saw the defeat of the Olympics proposals, it was certain that the antigrowth sentiment was to play a prominent role in the state's politics. In the same year, environmentally insensitive U.S. Representative Wayne Aspinall was defeated in his bid for reelection. In the summer of 1973, Colorado felt the Arab-induced energy shortages severely. The subsequent announcement of federal plans to undertake major oil shale development and to surface mine coal in the state became a dominant political issue. After more than five thousand acres of shale oil tracts had been leased early in 1974, State Representative Richard Lamm, a Democrat, mounted his gubernatorial campaign. His view was that "Interior and energy people in D.C. have more power in this state than I or our 100 legislators together."[32] The state needed a new assertive attitude toward growth management, he claimed. His election as governor and the election of a number of liberal Democrats to positions formerly held by conservative Republicans gave some credibility to his assessment of the state's sentiment toward uncontrolled growth.

Governor Lamm and his energy and natural resources advisors have generally favored conservation, and have made noticeable forays into national politics to benefit Colorado. The state set aside $73 million from royalties from its oil shale leases, using the funds to help impacted communities plan and finance facilities and services to meet the boom. The state's energy impact assistance office helps affected localities seek additional needed funds from federal sources. As yet the state has no strong land use plan or specific siting legislation, but Lamm has been a consistent leader of Western efforts to preserve the existing life-style of his state and the region.

The recent change in public attitude in Wyoming may be even more dramatic than that in Colorado. As one observer pointed out, Wyoming's former junior senator and a majority of the state legislature were development oriented and repeatedly welcomed coal industrialization as an economic stimulant. Former Senator Gale McGee hoped Wyoming would become the "Pennsylvania of the West,"[33] as the development proposal of the North Central Power Study was later to indicate it might. In June 1973, Governor Stanley Hathaway appealed to the nation's coal industry:

> Come look in Wyoming. . . . We have a lot of coal out there. . . . we'd like to have you. . . . And I always tell the people who are not particularly friendly to the mineral industries, if you want them all to leave, you'd better be prepared to dig pretty deeply in your pockets to replace this kind of revenue. . . . we will still mine millions of tons of coal in my state, hopefully find some more oil and gas and produce some more uranium. We could be ten years from now the energy state of this nation. I hope that we are. . . .[34]

Within a year of Hathaway's plea for massive mining and industrialization—a sentiment widely shared prior to mid 1974—Democrat Ed Herschler was campaigning on a platform of slow, controlled growth. Herschler ended a twelve-year Republican monopoly of the Wyoming statehouse. He was elected on a platform that stressed the need for strong state laws on industrialization and for halting the use of the state's aquifers for coal slurry pipelines to Arkansas. He was strongly committed to environmental preservation and development, only "on the state's terms."[35]

In the legislative session early in 1975, Wyoming passed three land use bills that gave the state some of the most comprehensive environmental management powers in the nation. The Land Use Planning Bill would zone the entire state according to uses such as agricultural, industrial, commercial, open space, recreation, and water storage. The state's water use law called for the planning, distribution, and use of water. The third significant piece of legislation, an Industrial Siting Act, required an industrial firm to justify site selection and to state the impacts of location and operation. These pieces of legislation would have been considered preposterous in the state eighteen months earlier, for land use planning is generally seen as antagonistic to the traditional frontier values of rights of ownership and use of land.

Colorado and Wyoming may have responded quickly to the threat of growth and of industrialization, but Montana has been the peer group leader in enacting strong laws to enhance the state's control over energy development. Upon the release of the North Central Power Study in October 1971, Montana realized the full threat to their state and region. Projections within the study indicated that eastern Montana's population might increase 500 percent almost immediately. Aware of the environmental consequences of this, Governor Thomas Judge, and especially former Lieutenant Governor Christiansen, began in early 1972 to appeal to federal agencies and industry for hard data on any plans to industrialize the state. They called for federal-state impact assessments. There was little response to the state's request until the Northern Great Plains Resource Program (NGPRP) was established. From mid 1973, the operations of the NGPRP were effective enough to provide Montana and neighboring states with needed information.

Montana's Environmental Policy Act of 1971 established the Environmental Quality Council to watchdog environmental problems, to recommend protective actions, and to supervise impact assessments. The Coal Task Force established in 1972 was designed to identify problems of coal development and to recommend necessary legislation. The spring 1973 legislative session was highlighted by the passing of a number of regulatory bills. The Strip Mining and Reclamation Act established an annual permit system and required sound restoration of surface-mined areas. In an effort to curb wasteful mining practices the Coal Conservation Act was enacted. Filing fees and the submission of impact studies were required of developers by the Utility Siting Act. Before water could be appropriated for new uses, a permit procedure had to be followed pursuant to the Water Use Act. Special taxes on the gross value of production of nonrenewable resource extracting industries were created through the Resources Indemnity Trust Act for long-range environmental improvement. In 1974 more environmental legislation was passed: the Severance Tax on coal which would render up to $60 million to the State Treasury in two years (the nation's highest rate), the Strip Mining Siting Act which gave the state authority to approve or disapprove the area of a proposed new mine prior to large-scale investment, and a moratorium on water use which suspended action on the processing of applications for large-volume extractions of water from the Yellowstone River Basin.[36]

Montana's endeavors to control the development of energy resources within its borders have been clear and certain. The posture toward the federal government and industry has been aggressive and defiant. Montana has publicly declared its suspicion that the federal government and industry will exploit and colonize it. The state intends that its environment will be protected. Montana's politics of environmental protection have been studied,[37] as has the nature of the economic and ecological motivations behind the strong legislative action in the state.[38] Apparently, much of the initiative for a multistate approach to problem solving in the energy-environmental issue and for a united stand against exploitation by the national government and industry has been generated by the state of Montana. The governor and his administration recognized that a united periphery of states would have more power than they would singly in bargaining with the center of power in Washington, D.C., and the East.

North Dakota also saw benefits to a united periphery of states. The state has played a seldom recognized role in facilitating interstate cooperation and information sharing in the five-state Old West Regional Commission subregion. Without the environmental protection banners of Colorado and the dramatic change in public opinion and legislative responses of Montana and Wyoming, North Dakota has yet been a positive force in bringing about periphery solidarity. From the beginning of his term in early 1973, Governor Arthur Link brought to the state a positive perspective on center-periphery relations. Link's statements on Project Independence in the fall of 1974 before the U.S. Federal Energy Administration clearly show his cooperative attitude toward development and yet his determination to preserve North Dakota's life-style and environment.[40] It has consistently been Link's position that his state would cooperate in energy development, but only if federal decision makers took North Dakota's counsel and needs into consideration when national energy policy was formed. The landmark memorandum of agreement signed between Link and the Bureau of Land Management in January 1975 points to the success of his approach to intergovernmental relations.

The model agreement to share information formally with the Bureau of Land Management has since been adopted by South Dakota, Wyoming, and Montana. North Dakota has consciously urged interstate cooperation in the Old West area, sharing research findings funded through Old West and, more recently, sponsoring through the National Science Foundation a series of four-state conferences attended by members of legislatures, state executive branches, and federal agencies. The state's Regional Environmental Assessment Program (REAP) is a model impact assessment research project that is unmatched in any state in the nation.[41] Decision-making capability in the state will be greatly enhanced when REAP becomes fully operational. A major motivation to operate the sophisticated REAP lies in North Dakota's intent to generate quality information. This is essential if and when the state is forced into the courts to protect itself against energy development which would degrade North Dakota's quality of life. There is considerable interest on the part of North Dakota's contiguous neighbors in REAP and much discussion of the model's utility on a multistate scale. Negotiations have begun to consider setting common standards for environmental quality indicators in the Northern Plains subregion.

The problems of energy development in Arizona are inextricably intertwined with the state's continuing bid for advanced economic development and the fact that Hopi and Navajo Indians own most of the state's fossil fuel resources. The state is traditionally preoccupied with economic stimulation, an objective which former Governor Raul Castro reaffirmed in his inaugural address of January 1975. In this speech he called for full cooperation with Rocky Mountain states in expressing an assertive, positive independence in dealing with the federal government.[42] This broad statement probably indirectly referred to the forthcoming regional energy policy organization, as it is clear that Arizona has expressed little concern for halting energy development or placing environmental issues higher in priority than growth.

Arizona has enacted almost no legislation directly concerned with energy development or the management of energy-related problems such as land use, water use, or the siting of power plants. Few serious environmental quality laws have been proposed. As most of the energy resources of the state are within

Indian lands, the state is not sure what right it has to develop or regulate them. Although the energy-environment dilemma has been questioned by some citizen activists in Tucson recently, the government of the state and apparently most of its citizens remain growth-oriented.[43]

Ranked almost last in the nation on the basis of annual per capita income, New Mexico is likely to continue its traditional drive to stimulate economic development. There has been some indication that citizens and some politicians in the Santa Fe and Albuquerque areas have begun to modify formerly progrowth stands. Governor Bruce King, in his earlier term of office, was devoted to economic progress and had limited sympathy for environmental causes. Former Governor Jerry Apodaca indicated a more moderate philosophy toward "development at any cost" than his predecessor. Although New Mexico passed an Environmental Quality Act in 1971, it was repealed in 1974 and failed to pass the legislature in 1975. Proposed bills for land use and energy conservation have failed to register significant support. Legislative efforts toward the control of uranium tailings and a proposed moratorium on high-level radioactive wastes have failed to win approval. Indian ownership of most of the coal and uranium in the state poses problems similar to those in Arizona regarding the uncertainty of state authority over coal development and regulation. Ranching interests have not been seriously affected by energy development as they have in Montana. If more energy reserves lay outside Indian lands and ranchers were more affected, one observer feels there would have been much more protest and organized opposition to projects such as the Southwest Power Complex.[44]

The state of Utah stands in greatest contrast to Montana, Colorado, and Wyoming. A survey by John McComb of the Sierra Club reported four times as many power plants planned for Utah as any other state in the region and no official opposition to the industrialization of the state.[45] Once the key issue regarding energy development in the state was the Kaiparowits power plant proposed for southern Utah. The plant was to be a linchpin of the Southwest energy network, the first of five mammoth coal-fired electric generating plants that would route most of their power to Los Angeles, Phoenix, and San Diego. In producing seven times as much power as is now produced in the entire state, the Kaiparowits plant and its network were likely to degrade surrounding national parks and recreation areas, and to alter drastically the socioeconomic makeup of a large area of rural Utah. Former Governor Calvin Rampton wanted the development for the jobs provided and the potentially great tax base.[46]

Energy development has had almost universal approval among state officials and the general public in Utah. In the eastern part of the state, where oil shale development is likely to affect both Utah's and Colorado's water and air and also induce some difficult local government problems, Utah shows far less concern for the environment than does its neighbor Colorado.

That the states view the various aspects of energy development differently is obvious. Even though the comprehensiveness of any classification of state positions to these problems will be limited, a general grouping of state positions on environmental protection may prove useful. Montana and Colorado, and Wyoming to a slightly lesser degree, have placed heavy emphasis on environmental protection. Moderately concerned is North Dakota, followed by South Dakota, then Nebraska. Less concerned would be New Mexico, Nevada, and Arizona in descending order, with Utah the least concerned about ecological standards.

Each of the states in the region appears eager to attract economic development, although those in the Southwest are perhaps most receptive. Montana and Colorado have had the most aggressive posture in the issue of federal-state relations. North Dakota and Wyoming have demonstrated preference for a less aggressive approach but, like each of the other states, continue to see the necessity for a multistate approach to center-periphery relations. This interest in addressing energy-related problems from a regional perspective developed noticeably during 1973 and 1974.

THE GROWTH OF MULTISTATE COMMUNICATION

Each of the states in the region realized that energy development projects such as Project Independence would affect its interests. This realization evolved rapidly during 1973 and 1974. Many of the states, represented by their governors or gubernatorial delegates, met frequently during these two years to discuss common energy-related problems, through such regional mechanisms as the Federation of Rocky Mountain States, Old West, and the Northern Great Plains Resource Program (NGPRP). Five of the governors—original members of WGREPO—and their administrative personnel in areas of energy policy and natural resources development gradually built a system of communications from their joint awareness of common problems. Nevada was not a member of any of the regional organizations listed above. Three of the four new governors elected at the end of 1974 (the governors of Colorado, Wyoming, and New Mexico) stepped into membership of one or more of these regional organizations and thus benefited from the interstate discussions of their predecessors. We will now demonstrate the existence of a growing regional consciousness during these years. The formation of the WGREPO was an extension of the regional negotiations already developed and also an indication that the existing regional organizations, as formed by 1975, did not suit the purposes of the governors.

The energy shortages, of petroleum in particular, of the summer of 1973 were severe in the Rocky Mountain states. Therefore it is not surprising that the ninth annual meeting of the Federation of Rocky Mountain States had as its theme the priorities and needs of the West in energy development. One of the major issues at this meeting was the West's need to have a national energy policy within which states could pursue their responsibilities.[47] It was resolved that the Federation would undertake energy and land use studies with regionwide application. An additional resolution of the member states was to endorse and support the development of the Kaiparowits power plant complex in Utah[48]—a development that Montana, Wyoming, and Colorado would probably not have endorsed so readily within their own borders eighteen months later.

The Federation's annual meeting in 1974 showed some skepticism about the ability of the federal government to produce a comprehensive energy policy. The governors were eager to strengthen their information base regarding energy development and its impacts, sought institutional mechanisms to facilitate research, planning, and regional policy making, and demanded a role in national decision making as it affected their states:

> The Mountain West must have some input to the development of a national energy policy. Our governors have made it abundantly clear that we do not intend to stand by in a colonial posture while the energy needs of other parts of the country are satisfied regardless of the consequences to the people of the Rocky Mountain States.[49]

The Federation immediately commended a coordination role that the West needed and undertook a study on energy development goals and concerns in the Rocky Mountain region.[50]

The Federation aided in communication between six states and played a positive le in the evolution of a regional approach to the West's energy problems. The Northern Great Plains Resource Program (NGPRP) has done likewise, bringing three more states into close contact. Member states of NGPRP were Montana, Wyoming, North Dakota, South Dakota, and Nebraska; the latter three were not Federation members. In contrast to the Federation, which moved to address energy issues after the summer of 1974, NGPRP had focused on the multistate implications of energy development since late 1972. In addition, NGPRP brought federal agencies, the administrations of the states and their governors, industry, environmentalists, agricultural interests, and Northern Plains Indians into joint research ventures and regular patterns of communication. NGPRP was a federally inspired venture but enhanced multistate communication.

The history of the creation of NGPRP and its operations provides a fascinating example of intergovernmental relations in the northern tier of Western states. It was sponsored by the U.S. Environmental Protection Agency, U.S. Department of Agriculture, and U.S. Department of the Interior with significant input and cooperation by the five states as well as local governments and special interests. NGPRP was a successful information transfer mechanism, commencing in the summer of 1972. Although Montana had urged a regional study in 1972 and was strongly supported by North Dakota by spring 1973, and although the regional office of the Bureau of Land Management in Billings, Montana, had asked its lead office for a multistate study in November 1971,[51] the states were at first suspicious that the NGPRP program was designed to help the federal government exploit energy resources.[52] Not until the program received the direction of John VanderWalker, some months after its faltering beginning, did it become productive and looked upon with enthusiasm and commitment by the states.[53] The product of the program, other than a comprehensive and informative inventory and impact assessment of energy development in the subregion, should also be weighed in terms of interstate cooperation and information flow.

Less prominently than NGPRP, Old West provided the northern states another institutional means for interstate information sharing and joint research studies on coal impact assessments. In the case of Montana and Wyoming, the Federation, NGPRP, and Old West each facilitated interstate relations. In other instances, states shared in memberships of Four Corners and the Federation, while others were brought together frequently by participation in both NGPRP and Old West.

These mechanisms fostered a regional awareness and action program, even though the NGPRP, Old West, and Four Corners had large federal backing. The topics of common concern multiplied throughout the period 1970–74 and stimulated multistate communication. The irresponsibly managed and irrational federal coal leasing program had been conducted with little or no state involvement and had angered each of the ten states in the region. The vibrations that spread from the severe boomtown experiences of Gillette and Rock Springs in Wyoming—as a result of unplanned and rapid ergy development—were grim lessons for each Western state.[54] The absence of opportunity for states to participate in federal decision making for the Southwest Energy Complex and North Central Power Study irked each state affected. In 1974 and throughout the years since, the issue of having sites for atomic waste thrust upon them by the U.S. Atomic Energy Commission raised strong opposition in several states. The energy park proposal for Montana also drew angry opposition within the state and from neighboring states. The predominance of federal and industrial input to oil shale leasing, at the near exclusion of state consultation, angered Colorado. Each of the Western states supported federal strip-mining legislation vetoed twice by President Ford. Each has uniformly resisted plans for federal preemption of state authority in matters of power plant siting, water use, environmental management, and rehabilitation of federal coal land. Finally, the statements of the states before the Federal Energy Administration in August 1974 indicated a mounting concern that there be state input in national energy policy.[55]

CONCLUSION

At the same time that the West continues to pursue the benefits of economic development, the states have grown increasingly aware in the last few years of the social and environmental costs of growth. One problem appears to be "people pollution"; some governors are distressed by the large migrations of Californians and Frost Belt indigenes into the region. The sparsely populated, rural life-style is a highly prized characteristic of Western culture. There is a general consensus among the states that some energy-induced industrialization is acceptable and desirable, but the extent of industrialization preferred varies among the states.

As a political region or section, the states of the West have never combined. Although they have shared common interests, especially in regard to economic stimulation, they have more often competed as individuals in the market for development than sought multistate cooperative approaches.

There are significant differences between the states in politics, tradition, and priorities, but concern for preservation of the Western life-style, agriculture, and tourism appears to be universal in the region. The growing importance of this concern and the recent barrage of proposals for exploiting the region's energy resources have acted as catalysts for change. The change has been expressed in unprecedented political developments in some states and a growing awareness of the need to adopt a unified front that will enhance the bargaining power of each state in dealings with the federal government. Common problems of public affairs and the operations of regional communication and research organizations have brought about interstate cooperation.

Emergent Conflict Over Energy Policies: Region Versus Nation? 4

Gubernatorial responses to the challenges introduced by rapid growth in the West became especially innovative late in 1974. The phases of political activity are easily recognized and are highlighted in the material following. First, the beginning of the politics of the creation of WGREPO took place between November 1974 and January 1975. Second, the phase of development involving extensive gubernatorial participation began 20 February 1975, the date of the official creation of WGREPO. In subsequent phases, the West's initially aggressive stance toward the East became somewhat subdued, and indications of federal suspicion grew. Finally, efforts at multistate policy formation in mid 1975 were marked by a phase of tension between the West and federal agencies, differences between states, and a major effort by the states to resolve their differences so that a regional organization could function effectively.

INITIATIVES FOR REGIONAL SOLIDARITY: THE ACTORS AND THE PROCESS

On 13 November 1974, Governor-elect Richard D. Lamm of Colorado met with John C. Sawhill in Denver. Sawhill had resigned the directorship of the Federal Energy Administration and was in Denver as a candidate for the presidency of the University of Colorado. Lamm was thinking of Sawhill as a prospective director of energy resource matters for the state. The following week, Lamm discussed his intentions regarding Sawhill with Governor Thomas L. Judge of Montana during a private meeting at the Democratic Governors' Conference in Hilton Head, South Carolina. Judge and Lamm were admirers of Sawhill's ability and his command of the technical details of the nation's energy problems; indeed, both Congress and the administration held great respect for him. During the meeting, Judge asked Lamm to consider forming an organization of some kind to include their two states and possibly other Western states, and hiring Sawhill or another energy expert to lead it. In Judge's view at that time, the organization should serve three purposes: bring expertise into the states' administrations at high levels; coordinate the activities of energy offices in the Western states; and influence federal energy decision making in Washington, D.C.[1] At a meeting on the final day of the Hilton Head conference, Judge and Lamm found their interest in the concept of a multistate or regional organization shared by newly elected Governors Ed Herschler of Wyoming and Jerry Apodaca of New Mexico.

Even though the ideas and suggestions for a regional organization were discussed between these four Western governors at Hilton Head, it was clear that Judge and Lamm were the principal initiators. Judge had been inaugurated in January 1973, having been elected on a platform that pledged strong strip mining controls in his state, a moratorium on water diversions for industrial use, and a "go-slow" policy on fossil fuel exploitation. The governor of Montana was convinced that, in regard to energy policy, the federal government was treating states' rights with contempt. Lamm had campaigned on a platform of assertiveness toward the energy industry and energy policy making at the federal level; he was nationally recognized as a proponent of growth management and as a leader sensitive to environmental problems. Both men believed they were being forced to build a political bloc strong enough to resist solutions to national energy problems that would lead to the degradation of the environment of their states. Each of them was certain that he would need more political power in Washington, D.C.—where the national energy policy is set—than any one state could muster. Accordingly, the tone of their arguments was set in the context of the assertion of states' rights. Equally important in their view was that federal energy policy recognize the need for effective strip mine reclamation, encourage conservation in energy consumption, and respect their determination to protect the quality of life in their states. What Lamm and Judge were endeavoring to bring about through a new alliance in regional politics was captured in the press. They did not intend to lock up their fossil fuels, but they did make the following statements. "Our position is easily misinterpreted," Lamm said. "We belong to the United States and we are proud of it. We are not blue-eyed Arabs trying to hold up this country. But we are saying that there are certain things that happen to colonies—whether they are Colorado or the Congo—if there is not some assertiveness on the part of their leaders. And we are not going to be colonized. We have seen what's happened to Kentucky and Tennessee and West Virginia and other states that have been the nation's coal bin, and we are not going to let that happen to us." Judge shared Lamm's sentiments and remarked, "If we are going to produce, it's going to be on our terms—not on terms somebody else dictates."[2] From early statements such as these it was apparent that states' rights would be a major issue of the organization and that federal energy authorities would have to deal with its contentions.

The momentum for establishing a regional energy policy organization gathered rapidly once the idea had been discussed in South Carolina. Judge and Lamm communicated frequently after Hilton Head and before the Western governors met in the first week of December in Kansas City at a Democratic governors' mini-convention. Judge met with Sawhill in Washington, D.C., late in November, at which time Sawhill confirmed his departure from the Federal Energy Administration and expressed interest in the regional energy office proposal. During the same visit to the capital, Governor Judge visited with Warren C. Wood, federal cochairman of the Old West Regional Commission (Old West), and discussed the possibility of funding the proposed regional organization through Wood's commission. Wood expressed his interest in the establishment of a regional energy office, but was not able to give Judge a commitment for the funds without first seeking the counsel of the U.S. Department of Commerce, the department under which Old West functioned. Wood then visited Sawhill on Judge's advice to discuss the concept of the organization. Feeling that he should also seek the guidance of the U.S. Department of the Interior, as Secretary Rogers C. B. Morton was the chief presidential energy adviser, Wood arranged for a meeting on 27 November 1974 in the office of deputy undersecretary William W. Lyons. In company with Lyons were Christopher Farrand, special assistant to Rogers Morton; Jack O. Horton, assistant secretary for water and power; and Alan B. Minier, counsel for Old West.

During this meeting at the Department of the Interior, Wood received mixed recommendations. There was a general lack of enthusiasm. The officials preferred to discuss the matter with Secretary Morton before making any statement, but they did express their individual opinions. Lyons was generally sympathetic to the idea of a Western states energy organization but did not wish to take a position at the time. Horton was favorable to the idea and thought the organization could serve useful purposes. Farrand, in contrast to the others, voiced opposition on two grounds. He felt, first, that the regional operation would be tainted if the states seeking federal funding could not themselves generate enough interest to support it financially. Second, he was suspicious of the motives of some newly elected governors who had criticized the federal energy program; Farrand feared that the Department of the Interior would be sued once a regional mechanism was established.[3] Secretary Morton was aware of the planning done by Judge and Lamm at Hilton Head, and Farrand further informed him of Wood's visit soon after the 27 November meeting. In Farrand's opinion, Secretary Morton "rather agreed with me"; Morton, he said, was prepared to adopt a wait-and-see posture with particular interest in what the organization might propose to do and who its staff might be.[4] Thus, even though there was a general lack of enthusiasm at the Department of the Interior, there was no evidence to suggest that Wood should veto the funding of Judge's and Lamm's embryonic organization.

Press announcements from Hilton Head of the governors' intentions to accelerate discussion of a possible regional energy organization probably also kindled the interest of the Federation of

Rocky Mountain States. The Federation, especially through the guidance of Executive Vice-President Michael H. Annison, had been developing a regional energy policy statement for six Rocky Mountain states since September 1974. On behalf of its members—Colorado, Utah, Montana, Wyoming, New Mexico, and Idaho—the Federation had undertaken the task of formulating a multistate energy policy statement for their consideration, assigning a team of staff and consultants to the task.[5] It was well known during the October–December 1974 period that the Federation was most interested in sponsoring—or at least in playing a leading role in—the operations of a multistate energy organization. During this same period, Warren Wood of Old West and Michael Annison of the Federation discussed the funding of the new organization and the roles of their respective offices in its future.[6] Wood told Annison that the Federation did indeed have a role to play in the future of a regional energy office serving the West, but his opinion was that it should not be the parent agency because the Federation received financial support from industry.[7] Annison and Wood were reported, by reliable sources, to be in conflict over the sponsorship issue. Annison, however, was handicapped by his inability to raise funds immediately as he would need assistance, in part, from his state members and this would in most cases have required legislative appropriations. Judge and Lamm were anxious to form the organization sooner than it would take legislatures to react. The Federation continued to work toward a regional energy policy statement in January, but when Old West agreed on 13 December to fund a regional energy office, and Four Corners likewise agreed on 29 January, the Federation lost its bid as sponsor.

By the early part of December 1974, it was apparent that Judge and Lamm were each actively promoting and seeking support for their objectives, and each appeared intent on obtaining official positions in the structure of the proposed organization. During this period Judge conferred informally with the other governors in the Old West region (Wyoming, North Dakota, South Dakota, Nebraska) and was accorded general interest in the proposal. Late in November, Stanley Womer, federal cochairman of the Four Corners Regional Commission (Colorado, Arizona, New Mexico, and Utah), visited the newly elected governors in his region—Raul H. Castro of Arizona,

Jerry Apodaca of New Mexico, and Lamm. During their discussion Lamm asked Womer if it would be possible to receive funding for a regional energy organization whose role, in part, would be to lobby and advocate on behalf of the West. Confirming the possibility of funding, Womer spoke of the utility of a nonprofit corporate structure but also stipulated that funding could not be used for lobbying purposes.[8] To this point, both Lamm and Judge favored establishing a lobbying representative in Washington, D.C., as well as an operating office in the West. By early December, Nevada Governor Mike O'Callaghan and Idaho Governor Cecil B. Andrus had indicated interest in sharing in the discussions of a regional organization. Involved would be the nine states in the combined Old West and Four Corners groups.

Developments continued to evolve rapidly. In the first days of December, Governor Judge announced that Western governors would discuss energy policy and the founding of a regional organization at the National Democratic Charter Convention in Kansas City on 7–8 December. Judge also made it known that he would be seeking funding for an energy office at the Old West meeting on 13 December in Lincoln, Nebraska. His statement revealed that, in his judgment, the Federation was expected to "go along" with Four Corners and Old West in preparation of a regional energy policy. Confirming the support of other states for his regional concept, he declared, "In addition to giving the energy-producing states of the West a strong voice in Washington, a joint energy policy would carry the weight of 18 'very influential members of the U.S. Senate and House.' "[9] During a Sunday meeting following the convention, interest was expressed by Utah Governor Calvin L. Rampton and Arizona Governor-elect Raul Castro, and the first collective action by the regional group of governors was taken. In a statement dated 7 December 1974, the governors of ten states (O'Callaghan, from the eleventh state, Nevada, was not present) sent the following message to President Ford.

> Dear Mr. President: The following Governors and Governors-elect urge you to sign S425, the Federal Strip Mine Reclamation Act. This legislation is critical to the people, land and resources of our states.[10]

Press reports on Monday, 9 December, drew attention to two

major issues of the Kansas City meeting that were to mark the governors' regional movement. First, the reports identified a divisiveness within the group, and second, they disclosed a sense of suspicion in some state and national political circles. The issue that created this suspicion involved an interest on the part of several governors in joint political action to influence the Democratic Party in its choice of a presidential candidate in 1976. These governors, unnamed, wished to push for protection of the environment as part of national energy policy in the Democratic platform. The effect of this issue was to create the suspicion that the regional energy organization would not confine itself to purely energy-related matters.[11] Associated with this was the suspicion, expressed to the author, on the part of certain administrative officials (and even governors) that Lamm or Judge might use the regional office as a springboard to national political office.[12]

The other issue raised in the press on 9 December had more immediate impact on the evolution of the regional energy office. This was the proposal to hire John C. Sawhill as the leader of the embryonic organization.[13] Several governors could not agree on Sawhill as leader, and their differences were aired soon after the Kansas City meeting.

In the first week in December, it was clear to a number of state and federal officials that Sawhill had the vigorous support of Lamm and Judge as the energy coordinator for the West. On 11 December, Lamm would neither confirm nor deny that Sawhill was being considered for such a job.[14] Why Lamm would not comment on Sawhill at this point is not certain except that he may have been waiting on the deliberations of neighboring governors. What is certain is that Sawhill forwarded a seven-page memorandum, dated 6 December, to Judge and Lamm, detailing his views on how the regional office should be funded, structured, and operated—as they had previously requested from him.

Before discussing the effects of the Sawhill candidacy upon environmental interests and upon certain Western governors, it is appropriate and necessary to review Sawhill's organizational recommendations. These suggestions were closely followed, consciously or subconsciously, when the states eventually established the Western Governors' Regional Energy Policy Office in

February 1975. In his memorandum to Judge and Lamm, Sawhill drew attention to the widespread concern among Western states for maintaining environmental quality in the face of federal and industrial pressures to accelerate the development of the region's energy resources. He felt that a regional approach was appropriate because of the existence of a multistate consensus on controlling growth and the rate of development. Other parts of the rationale for a unified Western approach were as follows. The states would require similar development-control policies, and each would find it necessary to adopt comparable legislation to stop industry from playing one state off against another. In addition, Sawhill argued that the low-population states needed a unified political front in order to influence key aspects of federal legislation. By "key aspects" he meant the federal provision of planning grants, front-end financing and funding of research, improved coal and oil shale leasing policies, and improved environmental regulations.

In Sawhill's view, the proposed energy office would have three functions: (1) the development of an energy policy that reflected regional interests; (2) the development of specific programs for states to consider in carrying out their responsibilities for controlling energy development; and (3) liaison with outside groups such as Congress, state energy officials, and industry. Further, Sawhill outlined the policy and communication activities that would be needed to fulfill these three functions. He demonstrated his preference for establishing the office in the form of a nonprofit corporation, taking funds from the Old West and Four Corners commissions, and having a close working relationship with the Federation of Rocky Mountain States. But he made it clear that the new office should be an entity distinct from the Federation because of the Federation's close association with industry. Staffing in Sawhill's estimate could be made up of about twelve in a regional city such as Denver, and two in Washington, D.C. The first year budget was almost $400,000.

The candidacy of Sawhill and the specific details of his organizational proposal were still, in the main, matters handled only by the two governors, Judge and Lamm. Judge used this proposal in his statement before a meeting of the Old West Regional Commission in Lincoln, Nebraska, on 13 December. Here, Judge argued forcefully for a multistate approach to energy policy formation.

> There is a need for states . . . to start getting together and trying to come up with some kind of policy instead of each state going its own way, completely fragmented. I think that we can lead the country, and I think one of the problems we have today is that we don't have any national energy policy. I think that if the region, the western states, gets together and comes up with a strong regional energy policy statement saying this is how we stand on these important issues, and aggressively goes after funds and projects that will benefit our region, I think it is going to be a very significant event for us and will benefit each of the states of the region.[15]

As his comments indicate, not only did Judge argue along the lines expressed in Sawhill's proposal but he was also sensitive to the leading role that the West might play in setting national energy policy, at the same time benefiting from material advantages. He declared that he had received favorable reactions from regional congressional figures who had responded to his inquiry about the advisability of establishing a ten-state energy policy.

During the meeting, Warren Wood made it known that the Department of Interior would welcome the states joining together. Governor Link of North Dakota seconded Judge's motion to have Old West and Four Corners authorize funds, each commission to appropriate an amount not exceeding $200,000. The motion was carried unanimously, although the record does not document the positions of Governors Link, Richard F. Kneip of South Dakota, and Stanley Hathaway of Wyoming, who had not been noticeably vocal at this point. Judge then notified the group that Governor Cecil Andrus of Idaho had offered $50,000 from his Pacific Northwest Regional Commission as a membership fee. Governor J. James Exon of Nebraska declared that the Western plan was an "excellent suggestion" and necessary to "improve the interests of our particular area" and to avoid "coming out second best, certainly with regard to the need of agriculture." The governors were aware by this time that the energy industry was attempting to pair states off one against the other; it threatened, if governors would not cooperate, to leave some states alone and to develop coal in others.

In the week following Old West's acceptance of the proposal

that it allocate up to $200,000 for the energy organization, Lamm and Judge both made headlines, indicating that each was continuing to play a leadership role. On Tuesday, 17 December 1974, Judge announced that Sawhill had been asked to head the organization and was considering the offer. Sawhill's memorandum was released the same day.[16] One day later, Lamm confirmed that he was "spearheading formation of a Rocky Mountain regional energy program and that outgoing federal energy chief John E. Sawhill probably will head it." Lamm said the governors wished to discuss the position with S. David Freeman, former head of the Ford Foundation Energy Policy Project, before making a final selection.[17] The governor-elect of Wyoming, Ed Herschler, was quoted as agreeing with Lamm's regional resource protection philosophy the day prior to Judge's and Lamm's announcements. Said Herschler "I don't think the greatest interests of Wyoming are guarded from across the Potomac. . . . Our attitude [the Western governors'] is going to be wait just a damn minute."[18] He claimed that the Western states were building a coalition to give them "political clout" in national policy making. These first documented accounts of Herschler's position on regional energy policy were probably similar to the sentiments he may have expressed earlier at Hilton Head.

Sawhill had discussed energy policy differences with President Ford on 18 October 1974, and the president made Sawhill's letter of resignation public four days later. Sawhill's last day in office as the administrator of the Federal Energy Administration was 16 December 1974.[19] His candidacy for the Office of Regional Energy Coordinator in Western States had been criticized by the Environmental Policy Center on 11 December. On 23 December, strong opposition came from the Environmental Defense Fund (EDF), a nationally respected public interest organization. The EDF criticism focused on three areas of policy: (1) Sawhill's promotion of large-scale Western energy development, (2) his efforts to prevent the passage of strong strip mine control legislation in Congress, and (3) his agreement that clean air standards should be weakened and that states should not have more stringent air quality standards than the federal government.[20] The Northern Plains Resource Council, an energetic and influential environmental group based in Montana, also criticized the proposal to hire Sawhill. The Council objected to Sawhill's support of an Energy Facility Siting Act that would permit federal agencies to override state siting decisions and to Sawhill's encouragement of nuclear energy development as a safe technology.[21] Yet, in opposing the selection of Sawhill, the environmental interests made it clear that they applauded the idea of a coordinated regional energy program emanating from the West.

Whereas the environmental interests generally disapproved of Sawhill's record on environmental and energy policy, Joseph Browder of the Environmental Policy Center had some confidence that Sawhill could succeed in the West because he, at least, faced up to the energy conservation issue and was "restrained compared to many other Nixon energy types."[22] Governors Judge and Lamm knew from the details of Sawhill's proposal (of 6 December) that he was sensitive to Western energy and environmental problems. They also viewed him as a heroic figure who had defied the official policy of the White House to speak his conscience. It is obvious, though, that full disclosure by environmentalists of Sawhill's record dampened the enthusiasm of Idaho Governor Andrus and several other governors.[23] Andrus was reported to have withdrawn from the multistate negotiations because of Judge's and Lamm's promotion of Sawhill.[24] While Sawhill's candidacy appeared to be on the wane due to disagreements between a number of the governors—as well as Sawhill's decision not to locate in the West—former U.S. Environmental Protection Agency (EPA) Administrator William D. Ruckelshaus arose as a strong contender.

Ruckelshaus had served as EPA's administrator from the agency's inception until May 1973. In September 1973, he was chosen by President Nixon as deputy attorney general and was fired shortly after in the "Saturday Night Massacre" for refusing to fire Special Watergate Prosecutor Archibald Cox. Judge interviewed Ruckelshaus at a breakfast meeting in Helena, and Lamm interviewed him in Denver. Both governors looked with great favor upon Ruckelshaus, but as a potential Washington-based legal counsel for the Western energy office. For like Sawhill, Ruckelshaus was not prepared to relocate in the West. Ruckelshaus was also regarded favorably by regional and national environmental groups as a prospective associate of Western governors.

For some Old West governors D. Michael Harvey also stood

as a strong contender for regional energy coordinator. Harvey, a staff member of the Minerals, Materials, and Fuels Subcommittee of the Senate Interior Committee,[25] had addressed thirty legislators from the five Old West states in early January 1975 on national trends in legislation pertaining to energy, land use, and the environment.[26] He declined to consider seriously the directorship of the embryonic regional organization, however, since he was satisfied in his then current position.

At the beginning of 1975, the driving force for establishing the regional organization still rested with Governors Judge and Lamm. Several governors may have objected to the promotion of Sawhill, but only Andrus spoke out in public. It seems also that some governors were distressed by the continued announcement of new events from the offices of Judge and Lamm before they themselves had been consulted.[27]

Up to this point, Sawhill's organization memorandum was still the most substantial evidence of what the governors had in mind. There had been no full meeting of the governors to discuss the formal objectives or the role of a regional office. Both Judge and Lamm had encouraged an adversary approach toward the federal government, but for the most part other governors had not formally expressed their views about objectives, roles, and structures. The major stimulus to move the governors toward unification on policy came as a result of the visit of the secretary of the interior to the West.

Press releases of the Hilton Head, Kansas City, and Lincoln, Nebraska, meetings and of the possibilities of Ruckelshaus or Sawhill heading the organization were subjects of considerable interest and speculation in the Republican administration.[28] Whether these activities were the direct, or only, cause of Rogers Morton's visit to Denver soon after the governors' intentions were broadcasted cannot be ascertained. But at the request of President Ford, the secretary of the interior did invite each of the region's governors to Denver to discuss energy policy, and it appears that this event triggered the first full meeting of the governors. For this and other reasons January 1975 was a significant month in the evolution of the regional organization.

The secretary of the interior, who at that time was also the leader of the National Energy Council, wrote to each of the eleven Western governors on 3 January inviting them to a discussion at the Brown Palace Hotel in Denver. The official purpose of the meeting was "to discuss our national energy policy and its regional ramifications. . . . I hope our meeting will contribute to closer consultations between federal and state entities in the development and implementation of energy programs." The specific issues Morton planned to raise were related to coal leasing policy, water for energy use, future federal-state consultation, land reclamation, and energy conservation.[29] In a follow-up letter, William L. Rogers, special assistant to Morton, stated that the secretary wished to describe "a proposed mechanism for continuing energy policy consultation between his office and the governors of the Rocky Mountain Region."[30] To the governors, these topics were certainly of prime interest, especially the proposals for state-federal consultation on energy policy matters.

Colorado Governor Lamm acted as host for the visiting governors. On 17 January he wrote to each of them suggesting that breakfast on 26 January be used for a discussion among them, allowing an hour and a half before conferring with Morton. He proposed that they adjourn to his office that afternoon to "discuss matters of mutual concern."[31] On this same day, Rogers Morton dispatched another letter to each of the governors drawing their attention to President Ford's new energy plan announced two days earlier. Morton wrote:

> . . . As you know, the President has just announced a new energy program which will require the expansion of domestic energy supplies, especially coal, oil, gas and nuclear: The President expects coal production to double by 1985, reaching a level of about 1,200 million tons annually. He has directed us to design a new coal leasing program for Federal lands, subject to the provisions of the National Environmental Policy Act. Our Denver meeting will be a timely opportunity to discuss the considerations which would have to be made before such a program can be implemented. Let me emphasize that I want *your input* [emphasis added] before I make any decisions with respect to a new coal leasing policy as it might affect your State.[32]

The President's most recent plan to achieve energy independence had major implications for the region. Those aspects of the plan expected to affect the region directly, although not to the full extent of the President's figurings, were: 250 major new coal mines; 150 major coal-fired power plants; 20 major new synthetic fuel plants; and proposals to amend (to weaken) the Clean Air Act, to modify (to weaken) proposed strip-mining regulations, and to expedite power plant siting procedures (to preempt state authority).[33] The expectation of doubling national coal production by 1985 meant the West's allocation would quadruple from 56 million tons to 250 million tons annually by that year.

In general, it was the secretary's position that in view of the national energy shortage, the federal government considered a vast increase of its coal and oil shale production in the West essential. The governors' position was that if the region was called upon to supply the energy demanded by the nation, then the people of the nation should share in the environmental, social, and economic costs of the development.

Morton made it clear that national energy interests would override any local, state, and regional concerns about social, environmental, and economic impacts.[34] In addition, he stated that the federal government and the Department of the Interior welcomed the formation of a regional energy policy organization. He charged the governors "to come up with specific recommendations," challenging the states to become involved institutionally in the energy program. At a press conference following the meeting, he addressed the following questions to journalists:

> . . . Let me ask you some questions. How mature are these [Western] states institutionally? Can they handle this energy development? Can they make it unnecessary for us to take them on in terms of protection of water, protection of environment and yet [achieve] development of the resource? Or is there to be a constant state of war between the feds and state interests?[35]

In addition to his challenge to the Western states, Morton claimed that most of the states had been dilatory in making difficult decisions about the energy-environment dilemma. This point was countered by a number of governors, who drew attention to the confusion in both the region and the nation owing to the lack of federal growth and energy policies.

Judge maintained that "until such core issues as population growth and insistence on unlimited economic growth are addressed, the nation and the world can only stagger from crisis to worsening crisis. We're very concerned about federal preemption of states' rights in such areas as site selection and relaxation of air and water standards."[36] Morton replied that while the federal government wished to accommodate state concerns, federal action on behalf of the public interest would be necessary if every state had veto powers and national policy was blocked. Postconference comments by Lamm acknowledged the federal preemptive power; confrontations with the federal government, thus, should only be used a a last resort, he stated.[37] Lamm continued to argue for the establishment of a regional bloc to ward off the "greater political and economic strength of the East." Herschler of Wyoming asked Morton if the federal government would preempt state action on matters of environmental and tax policies and land use planning. Morton answered that he hoped that such functions would be reserved to the states for the next twenty thousand years. Andrus of Idaho stated bluntly that he would resist federal preemption of state laws and had no intention of seeing the quality of life in Idaho deteriorate "in the race for ruin under which we are being told we must develop energy today."[38] The states' wariness of the federal government was openly admitted by Kneip of South Dakota, who stressed the need for state-federal agreements of cooperation.

A statement by Link of North Dakota, not considered significant at the time, came to symbolize the possibilities for effective communication between states and federal agencies in the region—a statement which may have been an inspiration to some governors who were not fully prepared to commit themselves to Judge's and Lamm's organization. Link spoke of a memorandum of agreement that he had signed with Edwin Zaidlitz, State Director of the U.S. Bureau of Land Management (BLM) in Billings, Montana, two days before the Brown Palace meeting. The agreement was the culmination of almost two years of negotiations between the state and the BLM. In effect, the agreement bound the two entities to a coordinated approach for the development and implementation of land use studies and management plans within North Dakota. The bureau and the state agreed to share data and personnel and to negotiate cooperative agreements between local agencies and quasi-governmental bodies whenever appropriate to the functions of either partner to the agreement.[39] At the time, this action was one of the most significant in the area of intergovernmental arrangements in the region. In light of the general combativeness between Morton and the group of Democratic governors, it took courage for Link to assert his confidence in gradual negotiation procedures, carried on in a low profile manner. Link was no less concerned about the energy-related and agricultural problems of his state and the region than other governors, but he did have preferences for specific methods of negotiation with the federal government—and these had proved successful.

DESIGNING A REGIONAL ORGANIZATION

As a result of this meeting, it was clear the federal government believed the nation's energy demand took priority over the will of state or region. This clarification possibly perpetuated some Westerners' fear of exploitation, as Apodaca had explained in the meeting with Morton. The federal government appeared willing to involve the West in future decision making, with Morton giving the impression of not wishing to preempt state authority unless he was forced to do so. His concession that a multistate mechanism might help in regional and national policy formation probably helped in the governors' decision to proceed.

When they met in private session on 26 January, Judge and Lamm were anxious to push ahead with the formation of the regional office. It became apparent, however, that some governors wanted more time to discuss the objectives, the structure, and particularly the roles of the proposed office. Further, several governors were reportedly against having Lamm and Judge, alone, make decisions. Accordingly, an ad hoc committee was formed, comprising Judge, Lamm, Exon, and Herschler, with Herschler serving as interim chairman for purposes of discussing the formation of the proposed regional organization.

In a memorandum sent on 28 January to all participating states and the state and federal cochairmen of Old West, Four Corners, and the Pacific Northwest Regional Commission,

Herschler defined the purposes of the ad hoc committee as: (1) to examine goals and objectives of a regional energy policy office, (2) to examine institutional arrangements that might achieve these goals, and (3) to suggest names to be considered for staffing key positions.[40] Herschler's plan was to compile draft recommendations in a meeting in Cheyenne on 7 February and to deliver them to each of the governors before the National Governors' Conference in Washington, D.C., on 20 February.

At this point, the proposed organization was referred to as the Regional Energy Policy Office. As a result of the Cheyenne meeting, organizational proposals were developed and guidelines for its formation drafted. The committee tentatively proposed Denver as the site for the new office; the office would have a board of directors of the eleven governors and would be a nonprofit corporation. The meeting also produced seven areas of concern to the region: federal coal leasing policy, federal impact assistance funds, energy conservation, federal preemption of state laws, interbasin diversion of water, research funds, and strip mine legislation.[41] Both Herschler and Lamm spoke of Sawhill and David Dominick, a former assistant administrator of the EPA, as potential directors of the new office.[42] For further discussion in Washington on 18–20 February, each participating state would draw up a list of major policy concerns. Warren Wood of Old West was asked to draft immediately a contract for funding, preliminary articles of incorporation, and bylaws. Wood's legal counsel, Alan B. Minier, spent the following three days on the preliminary drafts, and they were dispatched to each state on 11 February,[43] allowing the states time to review the contract, articles of incorporation, and bylaws before the Washington, D.C., meeting. In the draft the new organization was tentatively named the Western Governors' Regional Energy Policy Office (WGREPO), the title it was eventually to carry formally.

At this time, however, the states had not yet agreed on the roles the WGREPO would play, neither its relationship to other regional organizations such as the Federation, Old West, Four Corners, and the Northern Great Plains Resource Program (NGPRP), nor the structure of the new office. State officials in North Dakota felt there were specific functions that Old West and NGPRP would still be needed to fulfill once WGREPO was established. They were not sure how WGREPO would best relate to these existing mechanisms. Certain prospective roles for a mechanism such as WGREPO required partisanship, politics, and policy setting, but a need also existed for technical expertise in an effective regional mechanism.[44] Interim Chairman Herschler was sure the purpose of WGREPO was not "to come up with uniform laws for all eleven states in a race against the federal government."[45] The New Mexico delegates gave evidence of the unsettled structural form as of the meeting in Cheyenne, from their discussions of a proposed executive committee that was to comprise members representing three subregions in the West.[46] The idea of an office in the West and another in Washington, "to monitor the federal scene," was still current.[47] In a lengthy telegram to the participating states, Herschler summarized the meeting, laid plans for further discussion in Washington two weeks later, and announced an election of leaders of the Western and Washington offices to be held at the 18–20 February session of the Western Governors' Conference on Agriculture in Billings.[48] President Ford, the secretary of agriculture, and leading federal energy officials were scheduled on the agenda, which was designed to address the dilemma between energy development and the future of agriculture in the West.

Another regional action that took place prior to the Western Governors' Conference may have had profound significance for the establishment of a political bloc in the West. This was the formation of a caucus of congressmen and senators from the same eleven Western states.[49] In attendance at the first meeting held on 5 February were Senator Frank E. Moss, Democrat of Utah; Representative Morris K. Udall, Democrat of Arizona; Senator Clifford P. Hansen, Republican of Wyoming; Senator Jake Garn, Republican of Utah; Representative John Melcher, Democrat of Montana; Representative Gunn McKay, Democrat of Utah; and almost thirty staff representatives of other Western senators and congressman. At this meeting, Senator Moss was chosen as chairman and Representative Udall as secretary. The group agreed to proceed with organizational plans and to address energy-related problems for their region.[50] Moss and Udall spoke publicly of the overwhelming problems in the West and of the West's "real regional community of interest." In Senator Hansen's opinion, "the things that unite the Western states are far more numerous than the things that divide us."[51] Warren Wood and Alan Minier of Old West spoke of the formation of the governors' regional organization, and plans were laid for the two Western groups to meet later in February to discuss their common concerns.

Perhaps the only detraction from the growing unity of the governors in the region was the decision by Idaho Governor Andrus to withdraw from the regional negotiations. In a telegram dispatched on 11 February, Andrus pointed out that his state did "not share the common denominator of coal, oil, or shale deposits that mold the interests of the proposed member states together," and, furthermore, that funding through the Pacific Northwest Regional Commission was not plausible. From this point, the number of participating states was reduced from eleven to ten.

By mid February, Old West legal counsel Alan Minier had drafted a statement of purposes of WGREPO. These purposes, as framed by Minier, were to remain relatively unaltered in spite of later revisions by state officials. Those expressed were as follows:

1. to foster interaction and cooperation among the participating state governments with respect to energy-related issues;
2. to foster the enunciation, both within and outside the region, of common problems and policies of the participating states; and
3. to encourage the participating states to exercise their full authority in developing programs, including unified policies, criteria, standards, methods, and processes for dealing with prospective Western energy development.[52]

When at the Western Governors' Conference Colorado was agreed on as location for WGREPO, subsequent articles of incorporation drawn up in Denver added a fourth purpose: to facilitate the role of the participating states in working with the federal government on energy policies.[53]

The purposes set forth by Old West were a step in the design of the regional organization, and it is interesting to note the position of the sponsor with regard to them. Because of the uniqueness of this project—never before had Old West or Four Corners jointly funded a project or separately funded an incorporated entity—Warren Wood sought the counsel of his superiors in the U.S. Department of Commerce. Before the drafts were dispatched on 11 February, Alan Minier worked to have the "purposes" of the project approved. He obtained the approval of Frank S. M. Hodsoll, special assistant to the undersecretary. In the following week, the general counsel at the Department of Commerce objected to some language in the preliminary document, but the objection bore no implications for a change of policy. On the morning of the 18 February meeting of the governors' ad hoc committee in Washington, Wood and Minier were instructed by the Office of Regional Economic Commissions in the Department of Commerce to emphasize the relationship of the purposes of a regional organization to economic development in the West and thereby satisfy the wishes of the General Counsel. This suggests that the Commerce Department did not oppose the formation of the regional organization, thus supporting the previously released stance of Morton and the Department of Interior.

THE WEST'S CONCERN FOR ENERGY POLICY: WGREPO FORMED

On the night of 17 February, the staffs of each governor and Minier met to review the draft of Old West and state reactions to them. Some minor changes to the bylaws and articles of incorporation were adopted. Representatives were in favor of establishing a second line of directors who could represent governors in certain working sessions, but, as Minier had stressed in his original draft, policy-making power remained in the hands of the governors and only administrative authority could be delegated. An unintended outcome of this working session was some ill feeling caused by staff not permitting three representatives from the Federal Energy Administration (FEA) to attend the meeting. This apparent shunning of federal authorities was due in part to a previous ineffective working relationship with the FEA and the staff's displeasure with FEA's miniscule show of interest in state input to federal energy policy. Over the next two days the committee members and staff continued to review the draft materials and to arrive at a statement of joint concern that could be distributed by the governors as official regional policy.

The minutes of the meeting of 20 February spelled out for the first time some of the attitudes and intentions of the governors with regard to the structure of the regional organization. Questions posed by Governor Exon indicated an unwillingness to allow federal cochairmen of Old West and Four Corners policy-making status within the organization. Warren Wood stressed, in response, that WGREPO was not to be considered a federally sponsored program and that both he and Stan Womer of Four Corners totally endorsed it. Implying that they did not object to the exclusion, he reminded the governors that they had received the endorsements of Secretary of the Interior Morton and Secretary of Commerce Dent. The governors, in addition, voted to incorporate, to locate the office in Denver, to adopt the articles of incorporation and bylaws subject to amendments in a subsequent meeting, and to accept the proposal for funds drafted by Minier. By approving the corporation, the governors had, in effect, held the first meeting of WGREPO. It was here that the functions of WGREPO's director were approved by all the governors as follows:

1. to create forums for impact from state planners, industry, environmentalists, and consumers of energy policy;
2. to provide liaison with the Federal Executive Branch and Congress to interpret legislative trends and administrative actions for ten governors;
3. to marshal technical resources of the region to develop specific programs;
4. to provide liaison with state energy officials regarding common approaches to energy problems;
5. to act as a clearinghouse for information on energy problems and activities in the region;
6. to make periodic reports to governors on regional energy developments; and
7. other duties as assigned by the directors.[54]

At the suggestion of Lamm, Governors Apodaca, Judge, and Link were nominated as the screening committee for the selection of the director.

The choice of the inaugural chairman of the board of directors was, for most observers, a surprise. Kneip of South Dakota nominated Jerry Apodaca of New Mexico, and the nomination was seconded by Governor Exon of Nebraska. Kneip declared, "We have a lot of activity going on in Montana, Wyoming, North Dakota, Colorado . . . and it might add some balance in the first instance having a governor from further south." Lamm nominated Judge, but there was no second and no further nominations. Apodaca was approved by a slim majority in a secret ballot in "a very close vote," according to Interim Chairman Herschler.

There were many guesses about the political reasons for this outcome, but three in particular were often repeated to me. The first was that either Lamm or Judge would have been ideal figureheads had they not offended some of the governors during the early months of negotiation by neglecting to seek approval before making public statements about the proposed participation of various states in the organization. In addition, it was clear that their environmental stand was more radical than that of others. There was some suspicion, too, that they intended to use the office for personal political advancement. The second interpretation was that a number of governors preferred a more gradual, conciliatory, and cooperative approach to dealing with industry and federal agencies than had been suggested by the earlier aggressive stance; these governors sought a figurehead who would not upstage them with vigor or technical expertise. Not many of the governors knew the energy issue thoroughly, and they did not want to lose control to an expert. Finally, a common speculation was that the northern tier of states had lobbied for ex-governor William L. Guy of North Dakota as a prospective director and that some southern tier states had agreed to support Guy's nomination in exchange for the selection of one of their governors as chairman. As Rampton of Utah was too busy to accept and Castro of Arizona showed little interest, the young, energetic, and ambitious Apodaca of New Mexico apparently accepted the offer of support from the North. As the voting and lobbying were carried out secretly, the underlying facts will probably never be known, but each of the three scenarios described above appears to have some credibility.

At press conferences after the official formation of WGREPO, both Link and Apodaca stressed the intention of the office to help prepare the states for input into federal energy decision making, rules, and regulations.[55] Chairman Apodaca pointed out that the organization's states all had Democratic governors, and although their Senate delegations included eleven Democrats and nine Republicans, in his state at least, the representatives to Congress did not disagree on the policy of carefully controlled energy development. Apodaca was confident that as one voice the ten states could become a powerful political bloc. The governors generally agreed that a Washington lobbyist or representative was necessary, and the intention was to hire one along with a director in March. Moreover, it had become obvious also that the general consensus was that confrontational stands were not appropriate in future dealings with federal agencies. The more aggressive approach, preferred by Lamm and Judge in particular, had by this time been toned down by the governors as a group.

Rampton of Utah could not agree to the strong position taken by others against the continuation of federal coal leasing, as Utah is perhaps the most energy-production-oriented of the states in the West. He did, however, strongly urge the formation of policy that would request impact assistance funds from the federal government and agreed to "go along" with the first official policy or statement of concern drawn up by the group. The official Statement of Concern defined six issue areas.[56]

1. *Federal Coal Leasing Policy.* There should be a prohibition against the issuance of additional federal coal leases until a federal strip mining act is passed by the Congress and signed into law by the President.

   As federal coal leasing policies are developed, continued dialogue with the individual states is vitally needed.

2. *Impact Assistance Funds.* Since the demand for development of coal, oil shale, and all other minerals in the West is a result of national needs, then there is a corresponding national responsibility to insure adequate relief for environmental and socioeconomic impacts.

   This additional relief should be in the form of discretionary, front-end federal funds which will enable impacted communities to develop the necessary facilities prior to extensive population influxes.

3. *Energy Conservation.* Energy conservation must not be examined as only an alternative in gaining energy self-sufficiency, but rather as the cornerstone of each alternative examined.

Consequently, it is essential that the Federal Government establish long-range, uniform guidelines, and that under these guidelines each state be encouraged to develop its own rationale for a conservation ethic so that it is tuned to the specific needs of the individual states.

4. *Federal Preemption of States' Laws.* It must be the position of the states that they will not allow federal preemption of any laws which the states have the express right to develop and administer.

   The governors of these ten Western states are concerned about present federal proposals to preempt the states' authority to exercise control over the location of energy facilities and the authority to adopt and administer clean air and water standards.

5. *Research Funds.* Under the present federal policy we feel that adequate commitment and funding had not been provided for development of alternative sources of energy, demonstration projects, and energy conservation research.

   Along with this federal commitment and funding, it is also mandatory that the states have significant input to where and how these funds will be expended.

6. *Strip Mining Regulations.* The vast expansion of coal extraction and conversion is seen by the Federal Government as a keystone to eliminating the energy crisis. Because of this it is necessary to establish broad federal guidelines related to strip mine regulations. However, these guidelines must not preempt the individual needs of the states of site-specific legislation.[56]

This statement was the second formal indication of the West's common problems and its intention to take the cause to federal authorities on a united front. By now Lamm and Judge had lost some of their influence over the regional movement. Now the driving forces of activity moved to New Mexico, and Chairman Apodaca pushed the organization ahead.

REGIONAL AND NATIONAL INTERDEPENDENCE: STATE AND FEDERAL STANDS

Apodaca was now responsible for screening prospective candidates for the directorship and for setting up WGREPO. The establishment of the office was to be preceded by filing a contract proposal to Old West and Four Corners. In addition, the articles of incorporation and bylaws required review and completion. As Denver was the site selected for the head office of WGREPO, it was decided that Richard Freese, assistant to Governor Lamm, would be responsible for drafting and filing the articles of incorporation. Working from the office of Governor Apodaca, Ron Zee and Mark Michel would supervise the filing of the contract proposal and assist with administration. Before leaving Washington, D.C., the staff selection committee interviewed Dominick and Harvey, two prospective candidates for director.

Chairman Apodaca now became a central figure in the establishment of the office. On 11 March he dispatched letters and copies of the Statement of Concern to all Western congressmen and senators and to chairmen of energy-related committees in the House and Senate. In the letter he outlined the rationale for forming WGREPO to represent Western interests in Washington, D.C., and the intention of having the office staffed and operating by 1 May.[57] The majority of letters were answered, supporting the regional action of the governors. Senator Moss of the Western Conference proposed a meeting with WGREPO in Washington as soon as it could be arranged. House Minority Leader John F. Rhodes, Republican of Arizona, expressed his approval of WGREPO—especially the right of state input in federal energy decision making—and the provision of federal impact funding for communities placed under stress because of national energy programs.[58] Senators Lee Metcalf, Gary Hart, Pete V. Domenici, Paul Fannin, and Henry M. Jackson, and Congressman John Melcher, among others, welcomed the unified approach to addressing the problems posed by large-scale mining developments.

Senator Milton R. Young, Republican of North Dakota, who had recently defeated former governor Guy in a close Senate race, expressed the wish that the regional office be established, operated, and staffed on a nonpartisan political basis. Young was of the belief that the organization would make no meaningful contribution otherwise.[59] The Democrats running WGREPO probably thought little of Young's opinion. Perhaps Young wished not to see future electoral competition from Guy, who was now clearly supported by the governors as the director.

Five candidates for director were notified by Apodaca on 19 March that the selection would take place following personal interviews on 1 April in Billings, Montana. The candidates were now limited to Guy, Donald Cunningham of the Denver Research Institute, John G. VanderWalker of the Northern Great Plains Resource Program, Jack O. Horton of the Department of Interior, and Carl York, former presidential science adviser and then consultant with the Federation of Rocky Mountain States.

Besides newspaper editorials and political figures, other interests in the region commended the governors for adopting a unified front. The Rocky Mountain Center on Environment held a workshop in Montana on 1–2 March at which leading agricultural, civic, and environmental interests discussed alternatives to widespread coal development in the West. The workshop concluded that the region should choose agriculture as the mainstay of the region's economy and urged the WGREPO to adopt this alternative as a top priority. The conference defined the governors' regional effort "as a critical and essential means to direct the future of surface resources in the West" and suggested WGREPO serve as a regional resource office. The role of the office, according to the workshop participants should be "to supply the governors with technical, social, and environmental data as a decision-making base for directing their states and influencing federal policies," and the NGPRP (see chapter 3) should act as an important source of information to WGREPO.[60] The issues raised here were to be addressed in Billings on 1 April at the Western Governors' Conference on Agriculture. On 25 March Chairman Apodaca announced that he had called for a meeting of the Board of Directors of WGREPO on 1 April and intended to deliver a major address on energy and agriculture at the conference.

The events that unfolded at the conference, together with several statements, gave clues to the regional and federal positions on a number of energy and environmental policy issues. The choice of director of WGREPO was itself a significant indicator of the stand of the governors as a group, even though the individual motives of each governor cannot be known. Of principal importance at the conference were the address by Jack O. Horton representing the Department of Interior, the address of Apodaca representing the region's governors, and the compilation of a second statement of concern which was subsequently discussed with President Ford when the governors met him in San Francisco shortly after the conference.

Horton, assistant secretary of the interior, entitled his address, delivered to an audience of six hundred, "An Alliance of Interdependence." His theme was that man's technical systems were symbiotically connected with natural systems of air, water, and land, and that, to cope with this interdependence, systems of government themselves needed to be interdependent. In effect, he stressed the need for environmental management and intergovernmental relations in the development of a region where the greatest challenge was neither political nor economic, but involved the region's value system. Traditionally predominant in Western values had been a rural life-style and conservation of the environment. Horton contended that the region's resources must be developed, but, in effect, this development was paving the way for the states to join federal authorities in planning in an orderly manner.

To the question of further federal coal leasing, which the governors, other than Rampton, had protested, Horton replied, "The moratorium on coal leasing will be in effect until we arrive at a new leasing system that will permit maximum protection of the environment and national land use planning."[61] He continually stressed that the objective of Interior was to protect land and water resources and achieve total resource management. Using the term "new federalism," he stressed the wish of Interior to have state and local governments make decisions they were better qualified to make than were federal agencies. He referred to the memorandum of agreement between North Dakota and the Bureau of Land Management as a prototype mechanism for intergovernmental land planning. An "Operation Communication" was proposed that endorsed close ties between Interior and the governors and would lead to the establishment of regular meetings, as well as other communication mechanisms between regional and Interior authorities. It was suggested that the staff of NGPRP be directed as a source of technical expertise to support WGREPO.

The interdependence that Horton stressed as necessary between government levels would need to be founded, in his opinion, on three principles: first, a commitment on the part of the federal government that, in exchange for the resource development it asks of the West, it contribute to a substantial program which would understand and mitigate adverse impacts; second, a commitment on the part of the Western states to view their responsibilities as both regional and national; and third, a commitment on the part of both levels of government to approach energy development with enlightened and cooperative land use planning. The assistant secretary of the interior pointed out the negative consequences that would probably follow should the states or the federal authorities resort to confrontation, obstructionism, or problem solving in isolation.

Governor Judge and several different state administrators of natural resource and energy offices in some states were impressed with Horton. His address was well received, helping to spread the notion that he would be a leading contender for the directorship of WGREPO. However, Horton, a native of Wyoming, was also viewed suspiciously by other members of the state's congressional delegation who feared that he would run for political office after the WGREPO post, perhaps against Democrat Teno Roncalio. Others thought it unwise to select a Republican from the Ford administration to direct an organization for ten Democratic governors.

The address given by Chairman Apodaca was a comprehensive statement of the energy-related problems of the region and one of the first official statements of what Western governors sought in national policies related to energy, growth, and environmental issues. It also spoke of characteristics of life in the West that helped form a regional concept of the area: for example, widely dispersed populations, a desire to protect the "wide open spaces" and the rural way of life, the multistate energy deposits, generally lower per capita incomes than the national averages, and limited influence upon policy formation in Washington, D.C. At times Apodaca struck a determined note:

> Let there be no mistake— the West will not become an energy colony for the rest of the nation. We will not sacrifice our greatest assets—our blue skies and clear streams, our unblemished plains and mountains—to an endless national thirst for energy.[62]

Apodaca also declared that the states of the region wanted

more of the pie of federal energy research and development than they had been accustomed to. Apodaca, like Horton, stressed the necessity for long-range national guidelines and the need to respect planning for orderly growth. In a spirit similar to Horton's, he called for intergovernmental cooperation among states and between the federal and state levels.

The conference drew attention to the conflict in the region between forces promoting energy development and those protecting the viability of agriculture. That the agricultural theme, so powerful in the West, should enter into the affairs of the WGREPO so early in the organization's existence is not surprising. Of greatest concern to agriculturalists, and to the environmentalists who joined them in coalitions in some Northern Plains states, was water use. Utah determined that industrial use of water in that state was ranked last in priority after ecologically desirable stream flow, domestic consumption, and agricultural use.[63] Although other states had not ranked their priorities for water use, it was an unspoken tenet that agricultural use held priority over industry in most of the states. Conference attendees were reminded by agricultural interests that the sixteen Western states represented produce 78 percent of the nation's barley, 78 percent of the sugar beets, 63 percent of the fruits and nuts, 58 percent of the sheep, 51 percent of the vegetables, 48 percent of the wheat, and 35 percent of the cattle.[64] The final day of the conference was devoted to receiving testimony, from a wide range of interests, on energy and agriculture. The testimony was recorded and dispatched to congressional committees dealing with agriculture and energy and to state and federal officials in the region.

Another recipient of the conference findings was President Ford. On 1 April each of the region's governors (except Exon of Nebraska, who was absent), as well as the governors of Oregon and Washington, signed a second Statement of Concern and forwarded it to the president. On 5 April a group of governors hand-delivered another copy of the statement to President Ford in San Francisco, where Governor Judge spent more than half an hour with the president, elaborating the major points of the documents. The first five points concentrated on the energy-related policy questions that had been published in Washington, D.C., on 20 February, when WGREPO was formed. In addition, there were several points related to suppport and protection of various facets of the agricultural industry. Again, one of the main thrusts of the message was "that WGREPO be given an opportunity to develop policy positions on the important issues affecting our states *before* significant federal decisions are made."[65]

Agricultural interests represented at the Billings conference probably looked with favor upon the selection of William L. Guy as the director of WGREPO. Guy, a Democrat, had been governor of North Dakota for twelve years. He was a farmer and outstanding administrator experienced with regional mechanisms in areas of river basin planning and tourism. Having served as chairman of the National Governors' Conference, he, as mentioned earlier, had recently lost a close Senate election. Guy was popular with the Democratic governors in the Old West area. The vote for his selection as WGREPO director was at first eight out of ten. Judge of Montana respected Guy but favored Horton or VanderWalker, each of whom he felt had more technical proficiency and knowledge of energy issues and alternatives and would lead the governors when necessary. Lamm was quoted as saying he would have preferred a "technician, not a politician," but, like Judge, he voted for Guy to make the selection unanimous. There had been some thought that Lamm had avoided the Billings meeting once he was aware the governors had made up their minds to select Guy. Lamm was reported to have been angry about Guy's selection, but he later denied having purposely avoided the meeting. With the selection of Guy as director and Apodaca as chairman of WGREPO, both Lamm and Judge had been outmaneuvered by the larger group of governors.

There is little information available with which to evaluate Guy's selection for the key position. What is known is that many of the energy advisers of governors of Northern Plains states had urged the selection of VanderWalker, as he had directed the NGPRP studies and was well respected by both state and federal officials. Guy's selling points, on the other hand, were his experience in intergovernmental relations and his awareness of the governors' desire to set policy. Those who supported Guy as director but who also saw the need for technical expertise were reported to have considered VanderWalker as deputy. Guy's diplomatic, courteous approach to politics, combined with his breadth of experience in administration, seemed also to suit governors such as Link with a similar background of success in federal-state negotiations. Guy's name was circulated among some of the governors early in February, before the Washington meetings, and according to some sources, the votes for his selection had been organized well before 1 April. As the governors' choice, Guy was a successful politician and experienced administrator but not someone well versed, as other candidates had been, in national and regional energy problems.

During his candidacy for the U.S. Senate in 1974, Guy had called for a slowdown of energy development in North Dakota, sought a temporary moratorium on the state's approval of applications from energy companies for use of water, and agreed that environmental considerations on public land should prevail over profit making.[66] After his selection as director of WGREPO, Guy referred to the establishment of the organization as marking "the end of development at any price of resources in the West."[67] In a congratulatory letter to Guy, Chairman Apodaca stated that the governors had clearly favored him as the director, in part at least, because he would be responsive to their needs and desires.[68] Earlier, they had demonstrated their wish to be the sole authority in making policy statements; Guy had observed that principle. Guy saw his role as assisting in policy development and coordination between states.[69] He understood WGREPO to be a historic undertaking in state government and an experiment in regional unity; he was prepared to move cautiously but quickly to establish the office.

The months of April and May were taken up primarily with administrative delays. Progress was slow as Guy attempted to establish the office, hire staff, and keep the board of directors informed. He was responsible for selecting both staff and a Washington representative. In mid April, Guy visited Apodaca in Santa Fe to discuss administrative and staffing matters. The articles of incorporation were then being circulated for approval by the board of directors, but Ron Zee brought to Apodaca's attention that the contract funding proposal was still in the hands of Lamm's aides.[70] By this time Class B directors had been assigned, one to each governor, for purposes of working as alternates on energy policy matters with WGREPO. When Warren

Wood of Old West called Apodaca on 28 April and Alan Minier wrote on 1 May, each asked for minor alterations to the contract proposal. On 6 May they held up filing the proposal with the Department of Commerce while awaiting signatures from state cochairmen of Old West and Four Corners. Aides to Apodaca suspected deliberate obstructions were being set to delay funding. The federal government was stalling, they concluded. Federal Cochairman Wood objected to these allegations, countering that the states were poorly organized and had caused the delay themselves.[71] If it is true that the funding proposal was still in Lamm's office on 14 April, as Zee reported, and the final articles of incorporation were not filed in Denver until 23 April, the states had taken in excess of two months, from 20 February, to file the documents—evidence to confirm Wood's observation. Funds from Old West and Four Corners were finally handed to Guy and Apodaca on 19 May, enabling Guy to proceed with more assurance.

FORMING REGIONAL ENERGY POLICY

Three important meetings of the WGREPO were set in early May. The first, requested by Guy, was to be held in New Orleans one day prior to the National Governors' Conference on 8 June. The meeting was required to pass resolutions called for in the bylaws and articles of incorporation. Guy also proposed a meeting of the Class B directors on 7 July in Denver for purposes of drafting the first comprehensive energy policy statement. The official statement was scheduled for formulation on 28–29 July in Santa Fe when the board of directors would meet with officials of the Department of Interior and the Federal Energy Administration. In a memorandum of 3 June to the governors, Guy gave the following description of the attitude of the general public toward the lack of national energy policy, or at least regional policy.

> The public is looking desperately for some entity to begin a responsible approach to comprehensive energy policy. I believe Washington would be much relieved to see our major energy-producing states moving into an active role in assisting in energy policy formulation.[72]

The special meeting of the board of directors and Class B directors in New Orleans on 8 June accomplished the adoption and approval of more than a dozen resolutions dealing with legal and administrative matters. In addition, the status of Nevada was discussed but left undecided. As a member neither of Old West nor of Four Corners, Nevada Governor O'Callaghan pledged to accept the ruling of the group on how his financial contribution should be handled. The question of finance was also raised in connection with Guy's perceived need to increase staff salaries to make them comparable to industry and federal levels. The discussion which followed led to a major change in the structure of the organization. Judge moved, and Link seconded, that the $30,000 set aside for the salary of a Washington representative be deleted and used for salary increases. Originally a Washington counsel had been intended as an essential aspect of the regional organization, especially in the view of Judge and Lamm. Now it appeared that the governors were anxious to establish the Denver staff quickly, and that they preferred to deal with the question of a Washington counsel at a later point.

Another important decision made at the meeting concerned the release of information from WGREPO: only the governors and their Class B directors were to be recipients of news releases from WGREPO. Releases were to be marked with a specific release time so that the governors had equal opportuntiy to use them. Each governor was then to decide on further distribution of information.[73] Here was another indication that the governors insisted on full control of the organization.

The discussion of energy policy in New Orleans was preliminary only, and more formal deliberations were planned for the meeting of Class B directors in Denver on 7 July. The official regional energy policy announcements were scheduled for 28–29 July in Santa Fe. However, the preliminary discussions alone were illustrative of differences between the states. Lamm and Judge moved that the six positions adopted in the Statement of Concern related to strip mine regulations issued in Washington, D.C., in February be reaffirmed. All agreed, except O'Callaghan of Nevada, who abstained; it was the same resolution that Utah had objected to in Washington, D.C., on 20 February. Judge then moved, with the support of Kneip, that WGREPO affirm its support for the recently vetoed federal strip mining bill and that it urge Congress to override Ford's veto. This motion also passed, but again O'Callaghan abstained. Judge and Kneip moved that WGREPO support S393, the amendments to the Federal Mineral Leasing Act of 1920, and although the motion passed, O'Callaghan and Castro of Arizona abstained.[74] Finally all the governors agreed that agriculture should receive special consideration in energy conservation programs that might raise fuel costs to farmers. Class B directors were to take up the challenge of preliminary policy formation a month later in Denver.

Up to this point the differences between the governors on certain policy positions had seldom been made clear. Rampton of Utah, O'Callaghan of Nevada, and Castro of Arizona had been far less vocal than others on both the regional movement and environmental and agricultural matters. Rampton was quoted as saying a master plan for energy development in Utah and surrounding states was needed, that the state would share its energy resources in programs of responsible development, and that WGREPO's creation was a necessary effort toward planning.[75] Castro made few public announcements on energy policy or WGREPO, but he had stated that he was in favor of supplying the energy needs of the nation as long as the Western heritage and the existing and potential agricultural base of the region were not destroyed. He had also expressed an interest in providing strong state input to national policies in order not to "have to react to federal rules, regulations, and orders."[76] From the record though, it can be easily deduced that Arizona in particular and Utah as well were far more tolerant of rapid energy development than most of the other states.

In a *Newsweek* article, 5 May 1975, the regionalism that had made the formulation of energy policy in Congress difficult was interpreted as regional unity, as a form of state separatism.[77] In contrast to this, an observer in New Mexico focused on differences between the states in WGREPO. For example, in New Mexico a new tax on the generation of electricity discriminated against consumers in Arizona who imported the energy. Because of the tax, three Arizona utility firms filed suit against New Mexico. The observer was further aware that several Western states were actively competing for energy research and development projects that would bring economic stimulation to the recipient state. The tradition of differences, as well as the historical record of interstate competition, prompted the writer to fear "a more politically explosive subject locally than national colonization of the West."[78] Both positions, however, remain speculative.

Some of the differences between the states were identifiable by the different perspectives provided by Class B directors in Denver on 7 July. The rationale behind Guy's endeavor to formulate policy was that the region could not afford to wait for national energy policy to be set. He defined a comprehensive policy as "a plan of management for the production and consumption of energy with full inclusion of consideration and management of all associated aspects . . . such as protecting a citizen's rights, . . . environmental concerns, . . . energy conservation. . . ."[79] In an attempt to structure the discussion at the 7 July meeting, Guy had distributed a list of certain policy areas and programs of action his office might undertake.[80] The meeting addressed these policy areas, asking what approach might be taken in the process, and offered a range of duties and possible projects. The Class B directors also proposed policy areas that might be brought to the attention of newly appointed Secretary of the Department of the Interior Stanley Hathaway in late July at Santa Fe. Responses to the first two items on the agenda demonstrated the range of priorities within the region (see chapters 2 and 3). Concomitantly, there arose discussion once again on what the purpose of WGREPO should be.

From the first hours of the meeting, it was clear that particular states had their own ideas on approaches to policy formulation. Dan Bucks of South Dakota pressed for long-range analyses of the interstate effects of any one state developing its energy resources. Charles Metzger of North Dakota argued for the formation of policy that addressed the problems and alternatives for the region as a whole. James Warnock of Arizona was skeptical of attempts to set multistate policies when no state had even drawn up its own energy policy. Each state had its own bias toward energy development and the environment, he stated, and his needed energy production to boost its economy. Monaghan of Colorado, Governor Lamm's aide, urged the group to unify at all costs or have the federal government and industry pick them apart. In the opinion of Clark from Nevada, however, without national growth and energy policies there was little purpose in trying to establish regionwide policy. Peters of Nebraska declined to comment at all. Collins of Utah pointed out that there was no consensus in his state with regard to energy policy and expressed doubt of any unity among the states, declaring that "state lines cause awful problems in the West."

Toward the end of this segment of the meeting Wicks of Montana, supported by Freudenthal of Wyoming, pressed the view that the nation was waiting for the West to take a unified stand, and to take recommendations and guidelines to the federal government. Chairman Guy commended Wick's viewpoint and moved the members to an analysis of the major energy-related problem areas within their respective states.

Not only did the approaches to policy formation and ideas about policy content differ widely between the delegates, so did the individual problem areas of individual states. What was agreed on, though, was the common problem of institutional inadequacy in the region. Regional or multistate authorities for addressing and regulating impact problems and for helping reconcile the energy-environment dilemma were seen as necessary by Wicks of Montana. Monaghan described the lack of procedures within Colorado for reconciling energy-environment problems and elaborated on the difficulties of "entry" to federal decision making. Metzger carried on with Monaghan's argument and pointed out the lack of institutional structures or procedures to facilitate state input to federal agency decision processes. The lack of entry points to decision making in neighboring states was of great concern to Bucks of South Dakota. Fruedenthal agreed to the need for mechanisms for state-state and state-federal decision making but emphasized the difficulties states seem to face in generating data. He pointed out how in Wyoming federal and industrial entities had access to information in both quantity and quality which the state had not. It was essential, he contended, that research institutions supply states in the region with adequate and timely information.

Of equal concern to each state, and emphasized by Collins of Utah, was the need to press for federal funds for community-impact assistance. Peters of Nebraska singled out the problem of slurry pipelines planned to traverse his state from Western coal fields to Midwestern and Eastern consumer states. As problem areas specific to certain states were discussed in turn, Guy nominated three delegates to draw up position papers on these problems for the Santa Fe meeting with Secretary Hathaway. Zee of New Mexico was to address the impact assistance funds issue, Metzger the water use priorities issue, and Monaghan the coal leasing issue. Their respective governors would present the arguments in Santa Fe, on behalf of the West as a region. This exercise was one of the first multistate procedures for facing common problems, sharing information, and bringing the problem to the attention of federal authorities.

In the later part of the day's meeting, several delegates forcefully presented their views on the role of the WGREPO, advocating means for approaching federal authorities and actions which should be taken. Metzger spoke of the lack of response from the Department of the Interior to requests from the states for consideration in policy making; there was near-hopelessness in the gentle approach, he maintained, urging some form of threat to force responses. These arguments prompted Guy to give his personal view of the purposes of WGREPO and of the posture of states vis-à-vis the federal government:

> Do not face the federal government as an adversary. We should tell them truthfully that we will be forced to go to court to impede developments proposed by them if they do not try to meet our needs. If federal authorities do not help us plan, we may be forced to scrutinize such things as state water permits more severely.

Monaghan raised the question of WGREPO having a purely technical function or having some lobbying or advocacy role as well. To this Guy replied:

> This office is not an articulator of any position. But we need to listen to groups and forward the information to the governors. We will do no original research ourselves. . . . The governors do not want this office to be a spokesman. This office is not to be an adversary of the feds. That interpretation is false. . . .
>
> This office should help states enact laws the federal government refuses to do, such as for slurry pipelines and strip mining. There is no such power we have under the present constitutional set-up and existing rights . . . that we could carry out leadership in spite of the federal role, or lack of federal action.[81]

In the meeting's closing hour, Bill Peters, the delegate from Nebraska, reminded the group of the history of divisiveness between the states, a point that Collins from Utah had addressed earlier. Peters stated that currently almost every state in the region was trying jealously to entice the U.S. Energy Research and Development Administration (ERDA) into locating a solar research institute within its borders. He cautioned that if the states did not communicate with each other, they could well be outmaneuvered by another bloc of states in the race to attract ERDA's funds. It is interesting that this reminder came at the closing stages of a meeting that had moved gradually toward an awareness of the similarity of problems in many of the states; all appeared to agree on the advantages of a unified approach. The group clearly wanted to plan "joint-state" action rather than "regional" action philosophically conceived. It was equally clear, however, that WGREPO had not yet defined a sound sense of purpose.

At an important meeting of WGREPO and officials of the Department of the Interior in Santa Fe, at the end of July, the governors showed remarkable unity. They officially set energy policy for the region. For the first time they arranged priorities in a long list of projects that Director William L. Guy and his staff would undertake. Guy announced the urgency of some of these projects, adding that information on certain issues would be needed by January 1976 when several state legislatures would meet.

The governors agreed on eighteen of twenty energy policy resolutions. This, together with their plan to testify before Congress and the Ford administration, produced a political show of strength. The demand for states' rights in planning the future of the region was obvious and determined. Unhesitatingly, the majority of the states in the WGREPO were openly resisting one of the nation's greatest drives toward industrial development in history.

Some strong and aggressive language was apparently deleted from a few policy statements. And it appeared that representatives of the federal government were prepared to be more receptive to the demands of the states for consideration in setting national policy.

The "preamble" to the statement of official regional policy matters had been evolving as Guy met with WGREPO members in New Orleans and Denver and had been formulated with some guidance from the states. At the Class B directors' meeting on the first day of the conference, Dan Bucks of South Dakota argued that the document had a slight prodevelopment flavor, to which he objected, and called for more emphasis on energy conservation. The delegates from North Dakota and Colorado agreed with Bucks on the question of energy conservation and the latter was given the task of drafting a statement agreeable to all. Overall, the delegates' agreement with the message of the "preamble" was clear, and Arizona, Colorado, Nevada, and Utah formally commended Guy's material. The following excerpt from the preamble draws attention to the rationale of the West's actions:

> We ten Governors have formed the Western Governors' Regional Energy Policy Office under growing pressure to develop the fossil fuels of our region to supply the increasing amounts of energy to the nation under a rapidly shifting pattern of energy production. This shift is due to depletion of natural gas and crude oil reserves, price increases of imported oil, environmental protection requirements, and increasing demand.
>
> It has become obvious to our ten western states that we must move in concert to protect our citizens and our environment as we seek to provide our share of energy for the nation in a situation where comprehensive national energy policy is not yet fully stated or clearly defined.
>
> As we work together as energy producing and consuming states we are conscious of the problems which lie ahead. It will not always be possible for the leaders of ten states to have complete agreement on policy matters because of the uniqueness of each state, but it is important that we strive for action rather than impasse and that we seek consensus based on the most factual, current data and evidence available to us.
>
> In order to bring into sharper focus the long process of defining energy problems, evaluating evidence, and arriving at solutions, we the Governors of New Mexico, Arizona, Nevada, Colorado, Utah, Wyoming, Nebraska, South Dakota, North Dakota, and Montana adopt the following policies and positions. . . .[82]

Guy began the discussion of seventeen general policy areas by announcing the considerable revision they had undergone since the Denver meeting. "They form the fabric of the region's energy policy if accepted and hopefully they will contribute to federal and national policy setting," he stated. Guy also indicated that in several instances the statements of policy had been written in a general rather than specific tone. He was of the opinion that consensus of the Class B directors, and later the governors, was more important at this stage than sharply written policies. From the lengthy discussion that took place item by item, it was evident that Colorado, North Dakota, and South Dakota in particular had scrutinized the statement and were alert to needs of clarification or emphasis. Other delegates participated in the session from time to time, but there were no major objections to the content of policy statements until Arizona reacted negatively to the sixteenth point, which called for opposition to the issuance of additional federal coal leases until an adequate federal strip mining act was passed into law. Arizona's objection, in conformance with Governor Castro's position expressed in June in New Orleans, was based on the fear of the state's utility companies that their coal supplies would be curtailed. After delegates from North Dakota and South Dakota spoke in favor of the tone of the policy item, the delegate from Utah moved that the statement be amended to say federal laws should conform with those of the states. This motion was accepted and the policy item thereafter included no reference to blunt opposition to federal coal leasing. Arizona Class B director James Warnock successfully argued for the inclusion of an eighteenth policy item calling for the region to receive equitable consideration in any future energy allocation programs brought on by national fuel shortages.

Even though Director Guy had given his interpretation of the matter in Denver earlier in July, the posture the West should assume when dealing directly with the federal government remained an undecided issue. Monaghan of Colorado and Metzger of North Dakota had both argued for an aggressive and confrontational attitude during that Denver meeting, but Monaghan was now of the opinion, as was Governor Lamm, "we get off on the wrong foot putting ourselves in a confrontation stance." Metzger stressed that North Dakota had built up generally good relations with the Bureau of Land Management and Department of the Interior during 1974 with regard to assessment programs. And even though his personal wish was to be more aggressive, his state would prefer not to agree to an aggressive statement on the question of federal coal leasing. In the eighth policy area—the states' opposition to federal preemption of laws—Monaghan again argued against the "confrontational posture." This statement was reduced to call for joint federal-state exercising of responsiblities. These altered stands on the part of Colorado in particular were evidence of that state's willingness to tone down the confrontational posture that it had so long upheld. Similarly, the Class B directors appeared more willing to join federal authorities in negotiation than they had been previously.

The nonconfrontational theme arose again in the meeting of the board of directors. Guy began with the declaration that WGREPO would be an extension of the energy staffs of each state, then read the preamble. Two notable responses were from Kneip of South Dakota and Herschler of Wyoming. Kneip stressed the importance of a non-confrontational tone in the preamble; regional cooperation with federal authorities should be its essence. Herschler's primary concern was water use problems, and he insisted that any solution or decision be made by a state and federal partnership, definitely not by federal agencies alone.

As had been the case in the Class B directors' meeting, the question of the continuation of the moratorium on federal coal leasing caused difficulties. In 1972 the Department of the Interior had instituted the moratorium to prevent further speculation on coal reserves. Arizona's Castro and the substitute for Utah's Rampton objected to WGREPO's demand that the moratorium be continued until Interior developed a comprehensive development program. Nebraska's Exon likewise expressed concern as his state was a coal importer. South Dakota's Kneip argued that "an outright moratorium is a little too restrictive"; the tone of the policy item should urge federal agencies to hasten the implementation of an environmental impact statement as the lack of such was what troubled the states. In general the governors passed through the subject of federal coal-leasing policy briskly. The main stance of the states and region was revealed the following day in meetings with representatives of the

Department of Interior and at the closing session of the conference that followed.

Secretary of the Interior Stanley Hathaway and Frank Zarb, Sawhill's replacement at the Federal Energy Administration, were unable to attend the Santa Fe meeting with the governors, so three senior members of the Department of Interior were in attendance as federal representatives—Jack O. Horton, the former applicant for the directorship of WGREPO, William W. Lyons, and Jack Carlson. Lyons began his address by stating that Interior intended to institute regional environmental impact statements on all coal-related activities before lifting the moratorium on federal coal leasing in about one year.[83] This might have impressed the governors but not as much as four statements that concluded his message: "We will give the governors and states participation in our decision making. . . . we will share with you the triggering mechanisms and lead-times for intended developments. . . . we would like to have a continuing dialogue with Chairman Apodaca of WGREPO. . . . and our mission is to listen to the needs of the states." The needs of the states would be recognized, added Horton, and "we will work with the states and the region before and during coal leasing." He described the Energy Minerals Allocation Recommendation System (EMARS) program which would be the prime mechanism and procedure for coal development and for state leasing. Carlson continued the message of coordination: "We have asked the Bureau of Mines to work with your states. . . . we have already drafted a strip mining bill. . . . we've already prepared a position paper on front-end assistance for communities undergoing rapid growth. . . . our position paper on development of geothermal resources will be announced by the Department in the near future. . . . and we will carry back your messages to Washington." At least verbally, federal authorities had certainly shown a willingness to work cooperatively with the states.

Although many of the governors commented later on the obvious move toward conciliation by the federal representatives, they were not content to let the matter rest. "We are encouraged by sounds of input from us," said Judge of Montana, "but what role will the governors have in the final discussions regarding energy development in the region?" Lyons replied that very definitely ultimate authority rested with the secretary of the interior: "We are charged by law to carry out certain functions. We want input from the states, . . . but by law the final authority lies with the Secretary." The question of final authority on water use decisions brought Horton into the excited exchange. Said Horton: "We offer the states the first right of refusal on water permits submitted to us. . . . we do have some rights to water where we build dams, . . . but I hope that it will never be feds versus states on the issue of water." This was in concert with the address Horton had presented in Billings in April.

Judge, who had initiated the concept of a multistate organization, still had a tendency to lead the inquiries and statements of the governors, as was evident in this formal meeting. Lamm of Colorado was quick to follow Judge, both in the regional movement and the discussion with the federal authorities. "We are deeply concerned with the entry point of state input to federal decision making, especially the level and the time of entry," said Lamm. "In the synthetic fuels program, to my knowledge, there was no state input. . . . we were not consulted on leasing or the selecting of sites in the in situ oil shale development program. . . . we have not been consulted on the issue of front-end funding to impact communities. . . . geothermal leases were given in Colorado and the state was not formally consulted. . . . now the Energy Research and Development Administration talks of state input to its policies but I doubt it. . . . Look, we the states want more than consideration because it is generally too late." Lyons replied that the environmental impact statement process of the National Environmental Policy Act was an input point for states. He asked: "Are you saying you need input early in [decision] procedures? . . . at what point do you want to see us?" In reply, Lamm contended: "The feds and we are an investment partnership. The impact statement point of entry in decision making is too late. We are asking for partnership at the primary point of decision making." Carlson refuted Lamm on some of his claims but concluded: "Yes, you have a point, we'd better get more comment in on the geothermal question and some others."

The next issue of debate commenced with Herschler of Wyoming, who asked: "How much energy is required of Wyoming in say 1985 or 1990? Why give more coal leases if enough is in existing claims to handle this production?" This question led into a fiery exchange on planning and political ideology. "We are going to allocate tonnages of production per state," said Lyons, "but . . . now we are not." Carlson interjected, "We will not plan the West. . . . it smacks of socialism. . . . we need a process to help manage orderly development . . . and demand and supply market forces are the more desirable system." Once again the EMARS program was raised, this time by Lyons, as the desirable management mechanism and the procedure the Department would use in further coal leasing and development.

Carlson's suggestion that the federal government was not interested in planning brought four governors into the debate immediately. Lamm asserted strongly: "You *are* planning the West, more than the states or local governments. You plan where our agriculture exists, how much water is to be available for it, where new towns will grow and where old ones will die. Proposals now in existence for oil shale development in Colorado have more potential impact than all decisions in Colorado in the past fifty years." Herschler interjected: "We feel in many instances you are playing one state off against another. After you have decided what to do, then you involve the states." Kneip exclaimed: "We are concerned about the fallout from development in other states. What do you intend to do in your regional environmental impact statements about interstate impacts?" Lyons followed in the brisk debate: "We will need to improve our ways of telling you what we are going to do." Lamm countered: "We are asking more than that; we are asking substantive impact on decision making." Link expressed his wish that there not be an allocation of tonnage per state for coal production, but he also stressed that the states were desperately in need of some guidelines. As Link pointed out, the region's only guide to the pace and scope of development was the fivefold escalation of 250 million tons of coal per year by 1985 as called for by the president in his 15 January energy policy message to the nation. Lyons confirmed the estimate but could offer no other guidelines to assist in planning. Chairman Apodaca ended the debate with the federal officials with a clear reminder that "we want to be included in decisions early. We have expertise in these ten states that could be useful to federal agencies. We can provide ideas that are in addition to those carried forward by our congressional delegation. We feel William Guy has an excellent staff and will be our first line of communication. . . . we intend to push for our interests." The predominant thrust of the governors' arguments was that the periphery [the West] wanted more influence on the center's [federal] policy making. In addition, the governors were building political power to enable them to have such influence.

In the final meeting of the conference, the governors discussed the achievements of the WGREPO venture. There was a generally optimistic view that the states had come a long way toward a unified approach to energy problems in the region. There was also the consensus that federal agencies had begun to change their attitude toward the West. Governor Kneip of South Dakota summarized this feeling: "I think the federal bureaucracy has begun to change its attitude. Let us not be too negative in our dealings with them." Judge felt that "maybe we're turning the corner [in dealings with the federal government]. . . . Only a year ago Washington had shown no consideration of state attitudes." "The three visitors we had today are symbolic of Washington's more open attitude," stated Apodaca of New Mexico. Governor Castro of Arizona flatly denied that the dissensions on the part of his state and Nevada, on two policy questions out of twenty, weakened WGREPO: "We will not all agree on all instances, but mostly we will." Both Lamm and Judge felt all their expectations had been met, each confirming their preference for negotiation and cooperation with the federal authorities in the future. Lamm was still convinced, as he had been in 1974, that the states were entering a new era of assertiveness. The governors agreed that their policy resolutions would not be left on paper as they would themselves share in the task of presenting testimony, on behalf of themselves and WGREPO, in Washington, D.C. Whenever an issue affected their interests, they planned to work in close cooperation with the Western caucus, the group made up of the region's Congressional delegation.

CONCLUSION

From the examination of the politics of the creation of WGREPO it is apparent that not all its creators shared the same aspirations of regional unification. Some of the governors, Lamm and Judge especially, led the movement, while others such as

Castro of Arizona and O'Callaghan of Nevada were enticed into participating. The interaction between the participants had major implications for the structure, powers, and purposes of WGREPO.

The early months of creation were indicative of the originally aggressive stance the organization's initiators wanted the West to adopt toward the federal government. Later, as the governors began to formulate the structure and purposes of WGREPO, as well as the selection of its first chairman, a more subdued tone became the line of strategy. At first, Lamm and Judge wielded the major influence on plans for WGREPO's role, but evidence of the gathering influence of other governors coincided with the more subdued approach. This quieter stance could also be attributed in part to the hint of control of federal partners who provided the purse for WGREPO. That federal authorities may have tried to stall the establishment of WGREPO is unlikely. Most federal agencies and personnel were ambivalent toward the regional movements, though at times their initially casual response may have developed into some fear of the threat of strong Western opposition to federal policy.

For any one state the fundamental motivations to participate in WGREPO would appear to be the promise of some advantage or gain. The negotiations between the participants for purposes of insuring these gains affected the design and purposes of WGREPO. Arizona may have chosen to participate lest it lose something by not doing so; perhaps it would receive favors in return as well. Nevada desired to be wired to an information center and to keep in touch with some of its fuel suppliers. Fear of the pollution from up-wind neighboring states and the need to import fuel from them were some of the reasons for South Dakota's participation. The environmental preservation issue and the intentions to acquire political clout for use in federal negotiations were what primarily prompted Montana and Colorado to participate. While land and life-style preservation generated North Dakota's participation, it maintained its stance against the aggressive approach to federal authorities.

Whereas each participant had individual motivations for joining WGREPO, there were collective interests common to all of them. All shared the suspicion of federal authorities and sought to gain leverage in dealing with them. This suspicion also led to the decision to exclude federal representation on the board of directors of WGREPO. Most of the governors expressed the view that the West had been exploited before, and would be again unless political pressure could be assembled. That each governor was a Democrat was also a factor, as federal authorities were not only the center target but they were also Republican. The mix of these collective as well as individual interests was reflected in the organization of WGREPO.

The politics of creation attested to the decision of the governors to give WGREPO no formal powers. Each governor thus retained full authority, insuring that WGREPO would be a servant of the states. The politics underlying the selections of chairman and director would seem similarly to indicate this determination. The governors retained the sole right to use information derived by WGREPO, and there was provision made for the timing of its publication so that each governor was on an equal basis when news was released. The energy policy organization was designed as a voluntary mechanism with advisory and information-transfer roles. Regional policy guidelines put forward by the WGREPO might be used or could be ignored by each participant governor.

5 Organizing for Gubernatorial Effectiveness: Energy Development as Catalyst

After four years of continual gubernatorial activity, the West by 1979 had moved its innovations through and well beyond the Western Governors' Regional Energy Policy Office (WGREPO). The primary catalyst of accelerating energy projects and industrial development remained. As the institutional shortcomings of the states in the West became more evident, proposals for refinement resulted in the creation of the Western Governors' Policy Office (WESTPO). Efforts to increase the capacity of the states to deal with national energy plans became more sophisticated. Significant financial support for the WESTPO experiment was provided by the Intergovernmental Science and Public Technology Program of the National Science Foundation and by the Rockefeller Foundation, without which the governors' effort would have been more difficult to operationalize. The WGREPO experiment had been instructive; among other things, it became clear that regional energy policy discussions must involve problems of other natural resources including water and agriculture. Dealing with such crosscutting problems and harnessing the capacities of the numerous existing and competing regional organizations required major institutional reforms.

In this chapter, as in the previous chapter, phases of political activity are highlighted. The politics of creation of WESTPO, the effort to unite sixteen Western states, and the entwined managerial and political alignments that characterize this interval are examined.

The WGREPO had been established during mid 1975 and was embroiled in activities on behalf of its Western states for a full two years. Throughout this time, the states argued for the necessity of WGREPO, as uncoordinated federal energy proposals posed serious problems for the Rocky Mountain and Northern Plains area. Within a brief period after its creation, WGREPO supported numerous activities such as the effort headed by Governor Richard Lamm of Colorado to press for federal financial aid for energy-impacted states.[1] While immediate Western energy problems were closely watched over by Western states, there was also the emergence of potential conflict with other regions of the nation.

In December 1975, a group of Northeastern governors met to discuss common energy and economic problems in their region. Led by Governor Hugh Carey of New York, the group moved toward the establishment of the Coalition of Northeastern Governors (CONEG), for purposes of coordinating the political might of their region in Congress and the executive branch.[2] Both the energy-producing states of the West and the energy-consuming states of the Northeast were taking actions to enhance their positions in dealing with federal proposals that would markedly affect them.

The WGREPO states demanded that the Ford administration consult them in policy making—a stance which caused administration officials to see Western governors as energy-manipulating Arab sheiks.[3] In this context, and as part of a nationwide resurgence of state assertiveness in the federal system, the Western governors began to move forward in their institutional inventiveness.

Almost coincidental with the establishment of WGREPO, the National Governors' Conference (NGC) disaffiliated itself from the Council of State Governments. With strong initiative from the then chairman of the NGC, former Utah governor Calvin Rampton, the NGC membership resolved that states should be more assertive and should lobby in national affairs. In addition, there was widespread dissatisfaction with the fractured performance of national multistate executive branch associations purporting to speak for the nation's governors.[4] Heavily dependent on federal funds, many of the state executive branch associations were "captives" of the federal government and noticeably unresponsive to gubernatorial and NGC requests for innovation. A report to the NGC recommended that the proliferation of these executive branch agencies be brought under control to unify the position of the states on national issues.[5] The resurgence of state governments in this national context overflowed to the West, where Governor Rampton of Utah was a foundation member of WGREPO and the Federation of Rocky Mountain States.

A principal actor in the politics of creation of WGREPO, the Federation was no more sure of its future as a regional organization late in 1975 than it had been a year earlier. Its annual meeting in the fall of 1975 included discussion of the continuing need for the region to deal with energy problems and of the doubtful outlook on an institutional role for itself. WGREPO and the Federation overlapped in their energy research services to the member states, yet each was an asset to the region. It was clear that the governors favored WGREPO and were rather ambivalent about the Federation's future. While there was general gubernatorial dissatisfaction with industry's strong role in the Federation and, in particular, its heavy dependence on federal funding, Rampton, as one of the Federation's founders, was reluctant to contribute to its demise. Instead he initiated a major study in October 1975 for the official examination of financial alternatives for the future of the Federation. This, he felt, would add to the spirit of regional institutional innovation in the West. As part of the study, the Federation conducted an analysis not only of its financial futures but also, and more important, of multistate organizations (MSOs) in the region. Undertaken at a time when there was both regional gubernatorial and national interest in state assertiveness in policy making, the study had major implications for WGREPO, the Federation itself, and the member states' policy-making machinery.

During the NGC-initiated inventory of state memberships in national executive branch MSOs, the Western states became aware of the myriad associations purporting to represent them. In addition, the report that Rampton commissioned through the Federation revealed a proliferation of Western regional organizations and interstate compacts. The five Federation member states, Colorado, Montana, New Mexico, Utah, and Wyoming, found more than 130 Western regional organizations with which one or more of them was associated. Futher, the report indicated many undesirable features of the heavy reliance of the majority of these MSOs on federal agency agendas, timetables, and finance. In combination, the five-state MSO proliferation study and the Federation's financial alternatives report stressed that organizational proliferation and reliance on federal funds were inconsistent with the concept of state control and Western policy development.[6] Soon after these reports were released and discussed at a Federation meeting in Cheyenne, Wyoming, late in April 1976, they were submitted to Chairman Calvin Rampton for further action. While Rampton was personally committed to "maintain a regional capacity for policy making and program development" and pressed his colleagues in neighboring states to pursue "our collective efforts to promote the shared interests of the states in the region," he was uncertain as to what form of institutional arrangement should be adopted.[7] Considerable effort had gone into the Federation reports, and there was a strong consensus among the five governors that some regional institutional streamlining was overdue. It was the Colorado delegates, Governor Lamm especially, who were to add momentum to the developments, but the principal architect of the conceptual framework that guided the Western experiments was Philip Burgess of the aging Federation.

The executive director of the Federation during the MSO study, Burgess had accumulated extensive knowledge of intergovernmental relations in the Midwest and in Washington, D.C., before moving to Denver in the fall of 1975. During the year prior to his association with the Federation, he had provided analytical assistance to the study of the U.S. Office of Management and Budget on strengthening the policy management capacity of state and local governments.[8] His perceptive and stimulating address to the Federation at its annual meeting in October 1975 was a challenge to the inventiveness of Western governors. He charged them to move ahead with consolidating the interests of the states, to pursue regional aspirations, to allocate more of their resources to policy management, to improve their access to research and technology, and to find more effective ways to work with the private sector in the public interest.

It was Burgess's view that the Federation was a unique institution appropriate in a time of reassertion of state government. Although requiring improvements, the Federation possessed valuable characteristics which suited the needs of the Western states in their efforts to mobilize scarce talent and human resources.[9] With his executive order of 24 March 1976, Lamm appeared to adopt a number of Burgess's key concepts in the state of Colorado on policy management and the strengthening of gubernatorial management capacity.[10] Lamm established a new Policy Management Council, using Burgess's thoughts to aid in the executive coordination of interjurisdictional problems. Impressed with the revitalized staff work generated by Burgess at the Federation, Lamm and Rampton were instrumental in retaining Burgess's counsel and following through on recommendations delivered to Rampton in May 1976.

The spring of 1976 saw a surge of support for Lamm, an accelerating momentum provided by Burgess, by a Colorado delegation of executive officials, and by the persistent activities of Rampton, whose seniority, national prestige, and reputation for integrity gave extra legitimacy to the Western cause. Lamm had been criticized by other governors a year previous for his overly vigorous, assertive, and bold initiatives in creating WGREPO. This time Rampton was a principal cosponsor of Lamm's initiatives. The governors developed greater respect for Lamm's energy and foresight, and gave him their wholehearted support. In letters to Jerry Apodaca of New Mexico, Ed Herschler of Wyoming, Thomas Judge of Montana, and Lamm on 12 May, Rampton asked for views on a new innovative regional policy-making organization and the future of the Federation.

In response, each of the governors noted the need for an effective vehicle for interstate regional cooperation. Lamm noted the need for a regional mechanism, directed and controlled by the governors, involving the private sector in an advisory capacity and serving as an umbrella for existing regional organizations. Apodaca endorsed Lamm's view of the need for a visible connection between the region's state governments and the private sector. The continued joint participation of the public and private sectors was urged also by Judge. Herschler echoed the concern of his colleagues by referring to a range of problems with the Federation, the Western Governors' Conference, and intimate private sector relations in energy policy organizations. Like his colleagues, Herschler was most concerned about energy problems—"the most vital current issue facing our region."[11] Herschler and the others were eager to press forward. They would meet in the near future as Judge suggested, although they were worried by the thought of increased state dues required for any new organization and as yet there was no blueprint for action.

For nine months, Burgess discussed his concepts for innovations in multistate relations and for strengthening executive policy management in the West. In June 1976, it seemed to him that his suggestions were receiving attention too slowly to be effective, and he resolved to resign from the Federation, skeptical as he was about the likelihood of implementation. In a meeting in Governor Lamm's office with Roy Romer, treasurer of Colorado, and Jim Monaghan, senior adviser to Lamm, he declared his intention to leave the West. When called on by Lamm, Burgess argued that the proliferation of MSOs in the region was more serious than expected and that the West's gubernatorial and executive policy-making capacity was in dire need of streamlining. The group discussed what actions ought to be taken, concluded that the MSO study should be reexamined, and decided that Lamm would correspond immediately with Calvin Rampton to prompt further action.

Within a span of four days from the meeting, Lamm wrote to Rampton and set out guidelines for a new regional organization.[12] He flew to Utah to discuss the proposal with Rampton, and the pair planned to ask eight of the region's governors to meet the following month at the National Governors' Conference to discuss urgent business. The lessons learned from the five-state studies that Rampton had commissioned earlier and the conceptual contribution of Burgess were both very clear in Lamm's detailed letter to the Governor of Utah.

> . . . . In thinking about the organization of multi-state activities, I believe that stronger efforts should be made to combine into one organization the 10 States that presently comprise the two Title V regional commissions—i.e., the five "Old West" States (Montana, Nebraska, North Dakota, South Dakota, and Wyoming), and five "Four Corners" States (Arizona, Colorado, Nevada, New Mexico and Utah). This configuration is also coterminous with the States com-

prising the Western Governors' Regional Energy Policy Office and Western States' Land Commissioners and includes all the States comprising the Old West Commission, the Four Corners Commission, the proposed Rocky Mountain Institute for Policy Research, and all the States in the Missouri River Basin Commission. . . .

I believe that the kind of multi-state organization we need should have significant policy and program development functions. In this regard, I believe the organization should be constituted as an "umbrella" regional organization designed, in part, to streamline and rationalize our relationships with significant State Executive branch associations and interstate compacts. The "umbrella" concept can not only save the Governors considerable time and money in overseeing the program and policy development activities of the major regional organizations for which we are responsible; it can also focus our time and attention more clearly on *policy and program development* issues while the staff of the "umbrella" organization works with the staff of the other regional organizations to resolve management issues, to schedule back-to-back meetings, to develop common agenda and reporting formats, etc. This kind of process can have a significant impact on our ability to use existing multi-state staff capabilities in a more effective and integrated way and it can significantly help in leadership of existing regional organizations to bring important issues to our attention in a timely, actionable manner.

Therefore, for these and other reasons, I propose that we establish a task force composed of representatives designated by each of the Governors of the 10 States. I would further propose that the task force be mandated to create a plan and program for an "umbrella" multi-state mechanism to strengthen the policy management capabilities of member States—especially the State's chief executive officers.

In order to get a decision to proceed at the earliest possible date I suggest we consider a meeting among the principals, perhaps in Hershey, Pennsylvania, to coincide with the summer meeting of the National Governors' Conference. I would view the Hershey meeting as an occasion to reach a final decision on the appointment of the task force

> and the final formulation of its mandate. In this regard, I would hope we could agree that the charge to the task force is to create a plan and program for *implementing* a decision to create an umbrella regional organization. The report of the task force could then be voted up or down following its transmittal to the Governors. . . .

As they had a year earlier in South Carolina, the Western governors assembled in the East, this time in Hershey, Pennsylvania, to discuss their strategies for the West. Perhaps they needed the change of location; in the words of novelist Larry McMurtry, "The East encourages one to think hard and subtly. The West prompts one to imagine."[13] Chaired by Rampton, the 5 July meeting, with eight WGREPO governors in attendance, agreed to form a new ten-state Mountain-Plains Task Force and to assign the staff of the Federation to support it. The Task Force would examine the problems of regional MSOs that had been reviewed in previous studies, recommend options for consolidating and streamlining MSOs, set forth guidelines for a management and staffing plan, and submit a preliminary report in two months at the Western Governors' Conference in Jackson Hole, Wyoming. The visions of institutional reform that had been shared by these few governors and staff in the West had become the subject of hard thought and decision making in the East. There was considerable excitement in Hershey among the members of the Task Force. Many of them had built WGREPO and had shared more than a year and a half contributing to regional energy policies. This accumulated experience would be an asset, but the challenge of consolidating regional agencies in the face of bureaucratic inertia would prove a serious obstacle. As they left Hershey, Task Force members generally believed that any new regional institution would be established and operating in six months, before Chairman Rampton's term as Governor of Utah was up. Three years later they were still short of their original ambitions.

PHASE I: JULY–SEPTEMBER 1976

Even though the original ambitions and objectives of the Task Force had been only partially implemented three years later, significant progress was made in the brief period of the first phase. The first few meetings, leading up to the Jackson Hole preliminary report, were for the study and analysis of options. Principal actors, major issues, and basic concepts of organizational form were revealed. On the other hand, potential cleavages became apparent between the Task Force staff and therefore between some of the states. The West had seldom before had the advantages of such sophisticated and precise staff analysis and feedback. Rampton had asked Lamm's office to oversee the chairmanship of the Task Force so that continuity would be guaranteed even after his term as governor of Utah ended late in 1976. Lamm and others were eager to have Roy Romer, treasurer of Colorado and a former commissioner of agriculture, chair the group; Romer was a senior public official and a highly respected manager. The Task Force working sessions were management oriented, businesslike, and scrupulously prepared for and documented by the staff. In part this efficiency was credited to Roy Romer, who relentlessly focused on gaining "gubernatorial control" over proliferating MSOs. In part, the momentum was due to Burgess's preparation, conceptual guidance, and professional staff support. It was obvious that most of the governors' nominees to the Task Force were committed to institutional experimentation in an effort to further the managerial and political might of the states in the region, but there were differences in the degree of commitment of the states. Certain state delegations were more committed than others, and the hardhitting, potentially pervasive objectives of the Task Force brought other differences into view.

Governors at the Hershey meeting were unsure which organizational format to pursue. As in the days of the earlier WGREPO innovation, they groped for appropriate institutional arrangements, turning the analysis of options over to the staff in the Task Force. At this meeting Rampton, Lamm, and others declared that the Western Governors' Conference was "not doing an effective job," even though its chairman, O'Callaghan of Nevada, was in attendance. While this declaration may have caused some embarrassment to O'Callaghan, he in turn spoke angrily of the deviation of the Western Interstate Commission on Higher Education (WICHE) from its original mandate, thereby exemplifying problems from the lack of gubernatorial control over Western MSOs. Governor Kneip of South Dakota expressed distress that the Old West and Four Corners commissions might be abolished by a new regional organization. Rampton assured Kneip this would not be so, as the focus of the task force would be on state executive branch associations and interstate compacts, not federal-state bodies. Rampton went on to state that a new regional organization ought to include the ten WGREPO states because Idaho, Washington, California, and Oregon have problems different from the Rocky Mountain and Plains states. He suggested a new Western States Conference with the inclusion of Nebraska, South Dakota, and North Dakota from their Midwestern Governors' Conference. Stressing a variety of points and demonstrating varying degrees of commitment, the governors arrived at a broad consensus on the need for reform and the critical importance of a united Western political and organizational front. They were timid at the thought of creating another maverick bureaucracy and opted to select task force aides who had their "ears" and were "close" to themselves. Each state was tired of creating public agencies that became rather uncontrollable, independent entities.

The Task Force working sessions not only demonstrated a common commitment to regional action among the state representatives, many of whom had established WGREPO, but also revealed differences in perspective and priority. In extensive debates over regionalism and the appropriate grouping of states, it was generally agreed that WGREPO states shared significant commonalities. David Freudenthal of Wyoming, though, was apprehensive about creating a new organization without extensive analysis of the functions it would perform. Not differing fundamentally with their colleagues, Bob Hathaway of Arizona and Nick Franklin of New Mexico were eager to save money by institutional reform. As was indicated throughout the Task Force meetings, Bruce Arkell of Nevada was suspicious of the rationale behind the proposals for a new organization based on Federation staff and was philosophically opposed to formal bureaucratic arrangements that drifted away from gubernatorial control.[14] Bruce Nelson of Montana shared the view that no new regional mechanism would be supported by his governor until it was clear what purpose it had and until there was confidence in its performance. Governor Judge had recently ascended to the chairmanship of WGREPO and was pleased with the performance of the organization he had helped create a year previously. The five states that had undertaken inventories of the MSOs and national executive branch associations were more attuned to the need for prompt action than states such as North Dakota and Nebraska that had not. The issue of urgency was treated differently by each state, but it was clear that the task force had more in mind than a witch hunt of MSOs. As Jim Monaghan of Colorado stressed in the meetings, "the *opportunity* to take advantage of existing regional assets is more significant than the challenges presented by apparent breakdowns in existing MSO's."[15] There was consensus on the need for regional unity and institutional reform. The Task Force insisted it *not* be included in the formal agenda of the forthcoming Western Governors' Conference in Wyoming.[16] The WGC had been put on notice, and Arkell of Nevada now faced problems of institutional loyalty.

The WGC and eight other major regional organizations were the objects of special detailed analysis in a total of one hundred MSOs surveyed by the Task Force and staff prior to the Jackson Hole meeting of the Western governors. The WGC was shown to receive barely 10 percent of the dues paid by the member states to its parent organization, the Council of State Governments (CSG). Instead, the bulk of 90 percent of $332,000 annually supported other national CSG programs.[17] Through a comprehensive series of hearings, site visits to MSOs, and working papers, the task force identified executive branch associations, interstate compacts, and civic leadership associations in great numbers. The ten member states were paying dues for the partial upkeep of eighteen compacts, twelve energy organizations, twenty-three organizations for water, twenty organizations for environmental management, twenty-eight organizations for human resources, and sixteen organizations for agriculture. Only three were found to crosscut major functional areas. The Task Force inventory did not include Federal Regional Councils, Title V and Title II commissions, nor any state legislative organizations. This extraordinary pattern of fragmentation and severe duplication, for only the eight MSOs analyzed in the first review, cost the state budgets $1 million for dues in 1975, and a further $1 million for the travel of board members for business meetings. The combined total budget of these eight MSOs alone was found to be $16.5 million, and barely 5 percent was in state costs. As noted by the National Governors' Conference some years previously, "The key problem is that associations tend to grow and to be viewed by the members as the center of the universe. The Governors . . . need to keep a hand on what the associations are doing and how much they cost compared to their benefits."[18]

There was no doubt that the governors' MSOs were heavily dependent upon federal finance. Travel and board responsibilities were debilitating and unnecessarily expensive to the states, and there was no central coordination mechanism to avoid poorly planned meetings or to arrange back-to-back sessions.

As Roy Romer had stressed as chairman of the Task Force, it was difficult to capture the attention of governors for any length of time. Therefore the study and analysis options had to be precise and hard-hitting and to hammer at the issues of MSO *accountability*, *accessibility*, and *usefulness*. Such a report was delivered at the end of Phase I. Entitled *Regional Policy Management: A Framework for Decision*, it pinpointed the need to strengthen the accountability of MSOs, to reduce state financial outlays, to reduce fragmentation in state policy development, and to establish a comprehensive regional mechanism.[19] The staff support and quality of analysis in the report were unmatched by previous efforts in the evolution of Western regional MSOs; the document was as "hard-hitting" as Romer demanded. The governors were encouraged to introduce "sunset" provisions to make MSOs justify their existence annually. As part of the recommendation for streamlining and for functional consolidations, the report proposed that WGREPO, the Federation, and the Western Interstate Nuclear Board (WINB) be organized into one MSO as soon as possible. All of these guidelines and positions were referred to as "regional policy management functions" and were supported by ideas for sophisticated management details such as annual program budgeting and common office facilities.[20] The Task Force was convinced that energy, agriculture, water, human resources, and natural resources were the five major functional problem areas requiring state attention. No existing MSO appropriate for these tasks of integration and regional policy management existed, and the creation of a small, versatile regional staff was recommended.

With the study and analysis phase almost complete, the Task Force and the Western governors met at Jackson Hole, Wyoming, in conjunction with the annual WGC meeting. Seldom had the governors become involved in such detailed working sessions with staff as at Jackson Hole. There was general agreement that the task force had done "first rate" work, there was no argument with the proposition that more scrutiny of MSOs would be desirable, and there was the longstanding consensus that relations with the federal government were still less than satisfactory. The West needed a strong multistate mechanism, at least for energy policy making. The work of the Task Force was broadly endorsed by the governors. The Pacific Rim states—California, Oregon, Washington, Hawaii, and Alaska—were invited and chose to participate in the second phase of staff work. But the new alliance of sixteen states was far from secure. Tension had developed between certain MSOs and the Task Force, between those governors who were ambivalent to the WGC and its few fervent devotees, and between some members of the newly expanded Task Force.

Momentum for the Task Force gathered when the executive director of WICHE played into its hand by approaching the governors at Jackson Hole for an increase in dues. The detailed analysis of WICHE and eight other leading MSOs was in the hands of the governors while Dr. Phillip Sirotkin of WICHE made his request. With the evidence at hand, his presentation and appeal did not look good, and Rampton and Lamm complained angrily. They claimed that the original mandate of WICHE from the states required the work of one or two persons, but now it employed two hundred people and had a multimillion dollar budget.[21] The spawning of such bureaucratic organizations annoyed most of the governors present, and Rampton paid tribute to the value of Task Force analyses such as the material at hand.

As a result of the Task Force report, most of the governors became more sensitive to the assets inherent in good research staffs and of the inherent limitations in trying to share research needs with regional attorneys general, legislators, and lieutenant governors, the way the WGC would normally do through the Council of State Governments. But all was not well with the Task Force. When Governor Robert Straub of Oregon was introduced to Philip Burgess at Jackson Hole, he remarked, "Hullo Phil, you must be the person trying to destroy [unidentified] MSO. . . ." The Task Force was well aware before Jackson Hole that a number of MSOs, particularly those with Pacific Rim state members, were hostile to it. The Task Force invited these newly admitted states to participate, for defensive purposes at least. Keith Burns of Oregon and others were especially suspicious of the Task Force work, and together with Jerry Norris of the WGC they prompted the inclusion of the additional five states. Bruce Arkell, a principal aide to WGC chairman O'Callaghan and a Task Force member himself, is reported to have taken steps to ease the tensions of the Pacific Rim members and to encourage the sixteen-state alliance to work for the integrity of WGC.[22] And it was O'Callaghan who took the lead to slow down the Task Force.

The politics behind the maturing Task Force and the differences of opinion as regards the future of the WGC can be seen readily in the stands taken by the governors at Jackson Hole. The WGC was costing $17,000 a year in dues for Utah, leading Rampton to remark: "About the only thing nice I can say about that organization is that it doesn't cost much."[23] Lamm shared Rampton's view: "There isn't any sense in having a Western Governors' Conference as it now exists if the Task Force ideas are followed through."[24] Some newsmen interpreted the clash of opinions as a movement toward writing an epitaph for the WGC, but the WGC had its ardent supporters. Governors Dan Evans, Cecil Andrus, and Mike O'Callaghan, especially, praised the annual WGC as "a useful forum for quiet, behind-the-scenes give and take."[25] Governor Lamm agreed but showed his strong preference for the support of good staff analysis and the development of options on crucial issues. In a caustic retort, O'Callaghan defended the WGC, indirectly criticizing Lamm and his supporters for being underprepared to discuss issues at the annual WGC conference. He went further: "Governors who do not come prepared should not be spending taxpayers' money."[26]

Given the differing views, it was not surprising to find Lamm moving to block O'Callaghan's wish to bring the Task Force under the control of the San Francisco office of the Council of State Governments, and hence the WGC. With some concluding remarks that "every one of these organizations [MSOs] should have to restate its purpose and convince people that it was necessary," Governor Apodaca endorsed the continuation of the Task Force studies.[27] Such "sunset" activity would prove very difficult for Apodaca and others even though intentions were good. The governor of New Mexico was already a loyal chairman of the Education Commission of the States, and groups such as WICHE were well endowed with protective state legislators who were out of the reach of governors.[28] While the West was going through this comprehensive self-examination and reform, the East was following quickly in its footsteps.

PHASE II: SEPTEMBER–DECEMBER 1976

As the Task Force moved ahead to address solutions and consider the designing of a new regional organization for the West, activities and comparable initiatives in New England and the Midwest led to concern about the possibility of a war between regions and states. The West was intent on improving its capacity to cope with impending industrialization and economic growth. In contrast, the traditional urban and industrial regions of the "frost belt," now dependent on others for energy and raw materials, were intent on arresting the flow of capital and population, and hence political clout, to the Sun Belt states.[29] Editorials on regionalism holding the key to future political power in the United States were part of the literature on the proliferation of multistate power blocs.[30] The establishment of the New England Congressional Caucus and its Economic Research Office in Washington, D.C., was evidence of bipartisan promotion and research designed to boost a six-state region's unity and political influence. At about the time of the Jackson Hole conference, 204 congressmen from sixteen states formed the Northeast-Midwest Economic Advancement Coalition. This huge multistate effort had quickly followed the establishment of CONEG by the seven Middle Atlantic governors mentioned earlier. Perhaps in their vigorous and complex negotiations, Western governors and the Task Force may have briefly lost sight of the original energy policy motives for institutional reform that had dominated the WGREPO era. But the forceful reports of political activities in the frost belt region were reminders of the West's vulnerability in Congress and before the Executive branch. Key members of the Task Force, and most of the WGREPO governors, maintained the momentum necessary for the design of options for the West in time for a special meeting of the Task Force states on 10 December 1976.

In a brief ten-week period after Jackson Hole, the second phase of Task Force work was completed. The thrust of this work was to make final recommendations for reorganization of MSOs and the design of a new "umbrella" policy management organization which would mark the demise of the WGC. In addition, the reports of the staff strongly urged the synthesis of policies in the five major areas of energy, agriculture, water, human resources, and natural resources. This aimed-for synthesis alone indicated a major experiment in intergovernmental relations. The proposals for consolidation of MSOs within the five functional areas, the new concepts for policy management and "sunset" procedures, and uniform management guidelines for budgets and accounting were extensions of a major plan for innovation in state-state and federal-state relations. Guidance for the governors came in a sophisticated, lengthy, and provocative report intitled *Regional Policy Management: The Task Ahead.* It was an extraordinarily thorough and challenging guide, and a credit to the Task Force. Nonetheless, some governors were not completely informed of the full rationale and implications of the reforms being proposed. And several of them were absent from the 10 December meeting when the final Task Force reports were assessed and endorsed by those in attendance. Overall, the comprehensiveness of the reports, the rapid pace with which they were developed and reviewed, and the tensions that built up before Jackson Hole all contributed to lively political strategies and tactics leading up to the conclusion of this second phase.

Immediately following the Jackson Hole decision to expand the Task Force to sixteen states, the smaller original group of WGREPO governors met in Billings, Montana, to take stock of recent developments. Three of the founders of the Task Force—Nebraska, North Dakota, and South Dakota—had not been included in the Western Governors' Conference. While these three states generally supported the effort to unify the West and to cut down the proliferation of MSOs, the two Dakotas were reluctant to contribute to the demise of WGREPO.

Governor Lamm, chairman of the Task Force, had received a long, detailed letter from William Guy, WGREPO staff director, in which Guy made a case for the preservation of a modified WGREPO.[31] In Billings, Guy spoke of the difficulties of operating WGREPO on federal finance and timetables, but his main argument focused on his reluctance to see a regional organization that differed from his "ideal" of seven to thirteen members. He thought the proposed WGREPO/WINB consolidation would be feasible if the "consuming states of the Pacific Rim" were excluded. In what appeared to be an oblique defense of the integrity of his agency rather than anything else, Guy proposed a new three-man committee of WGREPO governors to examine possibilities for a new arrangement of energy MSOs. Task Force chairman Romer declined the offer, and as no governor challenged Romer's decision, WGREPO's fate was virtually sealed. Within a few weeks, William Guy announced his decision to resign. It was left to Governor Kneip to speak in support of WGREPO—in effect, for his friend William Guy—for the last time at the 10 December meeting, as Kneip and Governor Link of North Dakota were thought to have been the principals behind Guy's receiving the WGREPO post almost two years earlier.

There were no other formal meetings of the Task Force governors until 10 December, when the Pacific Northwest, Four Corners, and Old West regional commissions held a joint meeting. Leading the momentum for Phase II activity were Roy Romer, Philip Burgess, and especially Jim Monaghan of Colorado, who directed the energy reorganization subgroup, the most controversial item of business. Gubernatorial impetus in this phase of work was generated largely by Task Force chairman Lamm, by the newly elected chairman of the WGC, Jay Hammond of Alaska, and by Rampton of Utah. Hammond wrote to each WGC governor late in September, underscoring the importance of the preliminary report of the Task Force, urging the continuation of the expanded Task Force, and arguing the advisability of disaffiliating with the Council of State Governments.[32] The major Task Force charter came from Governors Lamm and Rampton on 8 October in a letter of instructions for "phase two" to Roy Romer and Philip Burgess. But it was the governors' delegates working in staff meetings throughout October and November who provided the best indicators of the progress and pitfalls of the Task Force.

In the First Task force meeting of the expanded group, Nebraska and North Dakota were absent but for the first time California was represented. The states were especially sensitive to the mounting national debate on regional economic imbalances. Within their own regional group, though, there was great diversity of opinion on the need for a new umbrella organization that Romer stressed as indispensable in the "implementation and orchestration" of their proposals. Realizing that MSOs had member states who had to become willing partners before control of MSOs was feasible, Romer and Monaghan introduced new concepts of regionalism. In an effort to accommodate the new states, they played down geographic factors and stressed "elasticity" in member participation. Monaghan tried to explain: "There need not be strict commonality among states for those states to work together." In contrast, Bruce Arkell of Nevada objected to any umbrella concept that was "political rather than administrative," and Susumo Ono of Hawaii tried to suggest a compromise in the form of an ad hoc managerial process. Romer held that "a continuous staff function is needed to help governors monitor regional programs and manage the development and implementation of regional policy." Monaghan agreed: "There is no mechanism to enforce a comparison of the interrelationships between Western water policy development and Western agricultural policy, and an ad hoc approach would not be able to accomplish this."[33]

On the general view that enforcement of gubernatorial wishes could come through budget controls and appointments to boards of directors, the Task Force found some common ground. It was evident, though, that Burns of Oregon was not comfortable with the concept of "regional organization," whereas Nick Franklin of New Mexico urged greater haste toward forming the "umbrella." David Freudenthal felt the reorganization and MSO consolidation plans were too optimistic, Jacobs of Washington was not happy with the proposal, and Joe Nagel of Idaho was afraid of creating another organizational "monster." Like his colleagues, Nagel argued, "We have to defend ourselves" in the face of the regional economic imbalances challenge, but he was not sure how best to do it. At best, the Pacific Rim states were hesitant and poorly informed on multistate cooperation and the concept of policy management. Colorado, Utah, and New Mexico were eager to press ahead.

Within a couple of weeks of the important 10 December meeting, there were further significant occurrences for the Task Force members. Urged along at a punishing pace by Romer and in spite of protest from at least Dan Garry of South Dakota, the group increased the pace and took major steps to examine possibilities for MSO consolidation and elimination. Governors Lamm and Rampton, at the suggestion of the Task Force, advised all states to freeze the funding of MSOs for 1977 until the final report was in hand in December. In addition, a team of private auditing consultants, Touche-Ross and Company, endorsed the feasibility of the Task Force's plans for streamlining the budgeting, reporting, and auditing of consolidated MSOs. When Jim Monaghan announced his subgroup's final plans for procedures to consolidate the Western Interstate Nuclear Board/WGREPO/Federation, the Task Force was met with numerous MSOs that protested and lobbied against the severity of forthcoming controls, all of which appeared to emanate primarily from Colorado. A powerful and well-entrenched Western Interstate Nuclear Board (WINB) was resisting consolidation. Monaghan, as the head of the energy subcommittee, had more difficult and unpopular responsibilities than any of the other five Task Force subcommittees on agriculture, water, human resources, education, and natural resources. As Patrick Pine of Oregon said, "What comes out of the energy subcommittee could well set the pattern for the others. . . . all the other subcommittees are deferring to this group because energy is so important."[34] With the hard-hitting consolidation and elimination plans drafted, with agreement on the need for a new umbrella organization eventually indicated by Wyoming's Freudenthal, Arizona's Hathaway, and Hawaii's Ono, the Task Force submitted to the governors seventy-eight recommendations and a blueprint for the umbrella.

The subgroup of states within the WGREPO group who provided the momentum, together with Burgess's staff, were somewhat aware of the fragility of the sixteen-state partnership during the second phase. Alaska, Hawaii, Idaho, Oregon, Washington, and California had missed the first phase of the Task Force, nor had they experienced the solidarity of the WGREPO experiment earlier. Nebraska, South Dakota, and North Dakota had been partial participants and observers in parts of the first phase and had not participated at Jackson Hole. George Gagnon of North Dakota had indicated his state might participate only in the agriculture and energy aspects of the new umbrella agency, and South Dakota was distressed about the demise of WGREPO. There was dissension over the impending recommendations to reconstitute and reform the WGC, and many MSOs had pressed for exceptions to the proposed Task Force controls.

As the governors' time and accessibility were limited, it was imperative that Task Force members brief their leaders fully before 10 December so that the meeting would be as productive as possible. The staff was well informed—more so than many of the governors. Dixie Lee Ray and Scott Matheson, recently elected as governors in Washington and Utah, were not well informed. Furthermore, some governors, such as O'Callaghan, preferred to "huddle" at governors meetings with a minimum of the staff analysis and preparation that Lamm, Judge, and others preferred. Thus, the styles of gubernatorial action were discussed by the staff in advance. Task Force members ensured that they would be seated strategically beside their governors at the 10 December working session, to encourage clarity and momentum in the deliberations. Generally the Task Force was optimistic, it was very well prepared, and it felt that it had a strong mandate from most governors.[35]

Cochaired by Lamm and Rampton, the 10 December meeting convened in Denver to pass judgment on the proposal xor creating a new Western Governors Policy Council (WEPC) (see fig. 2) with plans for full implementation on 1 July 1977.[36] With the lengthy volume *Regional Policy Management: The Task Ahead* in hand, the governors shared their second working session with the task force. They were advised to establish new policy offices for energy, water, agriculture, natural resources, education, and human resources. Within the new policy offices, guidelines for amalgamating MSOs, more specific guidelines for eliminating and reallocating state dues, and recommendations for establishing new management mechanisms were laid out. The recommended consolidations would save $350,000 a year in state dues and could lead to the elimination of several MSOs in the course of about two years through use of the "sunset" provision.[37] In other business, the Rocky Mountain Institute for Policy Research (RMIPR) was considered as a valuable addition to the policy

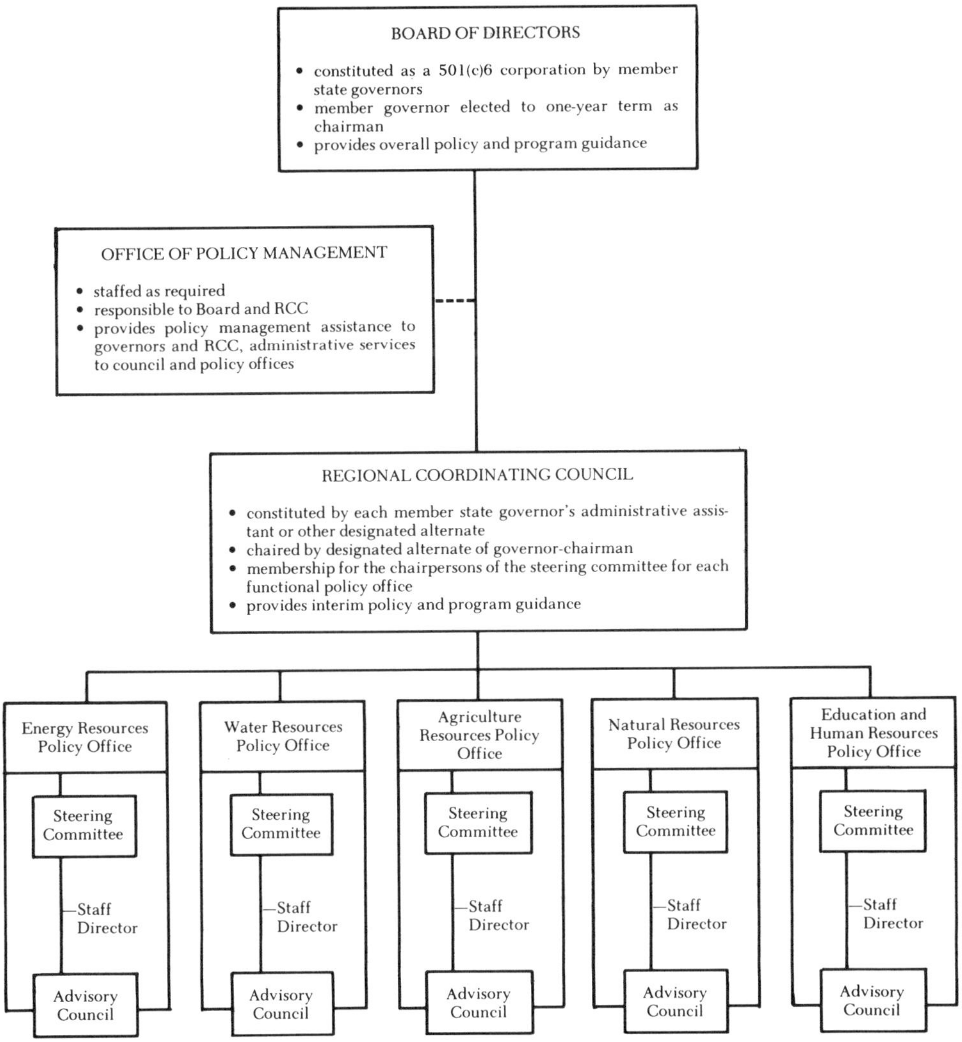

Figure 2. Organization of the Western Governors' Policy Council. Source: *Regional Policy Management: The Task Ahead* (Denver, Colo.: Western Governors' Task Force on Regional Policy Management)

management function of the WEPC. Finally, the new WEPC was to replace a reorganized and reconstituted Western Governors' Conference, giving the governors' new instrument freedom to lobby and promote activities on behalf of the states. All programs and activities of the new WEPC would be developed either through the functional policy offices or through the Regional Coordinating Council which would provide management supervision. Accordingly, the key elements of the new organization would be its five policy offices and the Regional Coordinating Council. A new policy management staff would provide secretarial services to the governors and their alternates.

The governors were prompted by the task force to reform significantly and to strengthen substantially their capabilities to conduct federal-state and multistate relations. With the exception of the governors of California, Washington, Alaska, Hawaii, and Oregon, who were absent, those in attendance unanimously endorsed all major proposals. There occurred some dissension by Kneip and Link once again over the demise of WGREPO. Kneip appeared to have arranged for William Guy to make another presentation. Guy declared: "Policy is for governors. . . . some of the organizations being considered are not for governors . . . but rather for their representatives. . . . you would not be willing to merge an organization with the National Governors' Conference just to save some money." Hitting out at the new superimposed structure of WEPC, Guy doubted that money would be saved in the two Dakotas and Nebraska. He voiced his opposition to a partnership with WINB and energy-producing states and set the stage for later complaints by Kneip.[38]

Governor Kneip rather angrily interrupted Monaghan's presentation on the proposed new Energy Resources Policy Office, disagreed with Monaghan's findings, and refused to make a decision on the matter, claiming a need for much more analysis for such an important decision. He felt South Dakota's voting power would be reduced by participating in a sixteen-state bloc as opposed to WGREPO's ten-state bloc. Idaho Governor Cecil Andrus helped to divert the temporary impasse by arguing: "It's the time and energy you give to an organization that gives you your voice, not the number of participants." This was promptly followed by Raul Castro of Arizona and Jerry Apodaca of New Mexico who each stressed that it was Kneip's "philosophical problems" and lack of delegated authority to his aide on the Task Force that prevented him from agreeing. While Kneip loyally defended his confidence in aide Dan Garry, the other governors pressed on for approval of the Task Force package of proposals. The "key vote" of approval to consolidate the energy agencies was a tensely awaited indicator to the Task Force that the governors were willing to "bite the bullet" and that there was hope of widespread support for other proposed reforms.

There were some differences of priority and perspective among some of the other governors as well, but they did not hold back approval of the proposals. Ed Herschler of Wyoming expressed uncertainty about the full implications of the Task Force report but was eager "to participate for the political strength that it will give Wyoming in federal-state relations." Governor Judge of Montana was not eager to welcome the Pacific Rim "energy-consuming" states into the WEPC but Rampton played the issue down and urged Judge not to be concerned. In contrast, O'Callaghan defended the nuclear technical capacity of WINB and expressed confidence in the views of the Pacific Rim group that had traditionally led WINB.

In an effort to clear the air of disagreement over the proper geographic grouping of states for WEPC, and in order to promote both compromise and recognition of significant differences between states, Philip Burgess expressed a timely and perceptive view of regional activity:

> The Task Force recommendations . . . are not designed to identify an identity of interest, but rather to invent a policy management process that can *forge* identities of interest, and to use different program management techniques to ensure that the proposed organization can work when a particular identity of interest is *smaller* than the *total* membership of states.[39]

Burgess had consistently been of the view that there were few genuinely "regional" problems among the Western states. Efforts to organize for the pursuit of "substantive" consensus were inherently difficult, he claimed. These efforts reduced WGREPO and other regional organizations to the lowest common denominator. There were common and shared problems, he went on, but the most vital was the lack of "institutional capacity" to handle policy management. Finding his statement correct, the governors endorsed the Task Force's advance to the third phase—implementation—although it appeared they did not fully understand the finer details of Burgess' concepts. The recognition of differences between states and the provision of new mechanisms for addressing them distinguished the WEPC proposal from the earlier WGREPO experiment. Yet, two years later, it was not certain that all the WEPC initiatives had settled into a comfortable routine between the Western states.

PHASE III: DECEMBER 1976–SEPTEMBER 1977

Observations by national press correspondents early in 1977 accurately captured the larger national context surrounding the activities of the Western governors, acknowledging the gestures to regionalism and political unity identified earlier at Jackson Hole. Generally, the conclusions of the 10 December meeting on organizational reform were interpreted as a continuing effort to grapple with energy policy in Western states. Gubernatorial initiatives were thought to be efforts in cooperative federalism that were recommended in a RAND report on Western energy development. From a distance, observers saw in the WEPC the well-known initiatives of Rampton and Washington Governor Evans to increase the clout of states in national affairs.[40] But few observers recognized what Neal R. Peirce saw: "Not even New England, the best organized region in the nation up to now, has been willing to take such tough actions, which inevitably offend entrenched bureaucracies and special interests."[41] Within the Western states, there was the overriding question of whether the entrenched bureaucracies and their political supporters would be willing to undergo consolidation and extensive supervision. Some thought the energy organizations were willing to accept consolidation, but this was a superficial and premature judgment.[42]

There was a complex political and organizational process under way within the West, but few examined its subtleties. The energy-consuming, nuclear-oriented Far West had little in common with the coal-producing Mountain and Plains states. As Governor Link had asked of the press after 10 December, "How do you have producers and consumers in the same room?" Patrick Pine of Oregon, after the same meeting, from which coastal governors were absent, remarked, "There are a lot of unresolved questions about how this will work . . . on the mechanisms these governors will have to be in there." While coastal states were concerned about MSO proliferation, Pine maintained that they were not eager to make their Northwest Energy Policy Project (Oregon, Washington, and Idaho) subservient to Rocky Mountain energy priorities.

It was not simply the contrasting policy priorities between groups of "producer" and "consumer" states that set the stage for political fragmentation in the third phase of Task Force work. There were also struggles over the form of the umbrella organization, particularly over the roles and staffing of the Regional Coordinating Council and Office of Policy Management (see fig. 2). This phase of multistate negotiation was afflicted with the continuing conflict over the proposed integration of WINB with the Federation and WGREPO to form a new Western Interstate Energy Board (WIEB). The establishment of WIEB (discussed later in this chapter) was to be the principal challenge during the nine months before the Alaskan conference of governors in September 1977 and for some time thereafter. Not only was energy policy the main catalyst for reform, it was also the first subject addressed by WEPC, the policy area in which the entrenched bureaucracies posed such a threat to success. Without success in the reorganization of energy MSOs the WEPC would itself likely fail. In addition, the Western Governors' Conference challenged WEPC in the third phase even though the 10 December meeting had voted to reorient and reconstitute the WGC. Facing such difficult political and managerial problems, the WEPC required constant and timely rallies of support from a variety of vociferous Task Force supporters.

The Task Force was divided over numerous issues, mirroring state positions and preferences. It was a fascinating microcosm of interstate conflict and cooperation. Detractors were predominantly from the Pacific Rim states and Nevada; they were often absent from Task Force sessions and at times infuriated their colleagues with early departures from meetings.[43] Task Force Chairman Romer and Staff Director Burgess had provided critical momentum throughout 1976. As their clout and credibility ebbed and flowed in 1977, critical surges of enthusiasm and leadership came forward from Bill Gordon of Alaska, Bruce Nelson of Montana, Kent Briggs of Utah, Bob Hathaway of Arizona, Nick Franklin of New Mexico, and especially Jim Monaghan of Colorado, who moved remorselessly and forcefully to lead in the crucial consolidation of the energy agencies.

In fall 1976 Jim Monaghan and Bob Hall began their leadership of the Task Force's Interim Steering Committee (ISC) on energy MSO consolidation and the characteristics of WINB with a background paper for the 10 December meeting.[44] WINB was the operating agency and administrative vehicle for implementing the Western Interstate Nuclear Compact, enacted among twelve states in 1969. Its work focused almost exclusively on technical investigations related to the development of nuclear energy in the West. Major recommendations were that WINB's functions should be encompassed by, and integrated into, the proposed Energy Resources Policy Office, that governors should invest in one person the duty of serving on the board of the new amalgamated energy agency, and that the existing $180,000 annual appropriation of states to WINB should be used to fund programs recommended through the new WEPC.[45] These recommendations would not only bind WINB into a locational consolidation. They interfered with states' representation on its board of directors, they appeared to challenge the integrity of the compact, and they made projects subject to direct supervision by a new governors' umbrella agency. Suddenly WINB became the object of widespread gubernatorial scrutiny. There was a rush to capture its dues and reorient its activities, and the problems of joint legislative-executive control underpinned a bitter struggle between the reformers and the defenders of WINB. The reformers, in particular Colorado's Monaghan, incurred the lasting wrath of most of the Pacific Rim delegates.

Late in January 1977, New Mexico's Franklin, Alaska's Gordon, and Arizona's Hathaway were eager to have the energy consolidation decisions returned from a subgroup of states to the full Task Force so as to "nip in the bud any suspicion that a few states were directing the energy consolidation." Nevada, Washington, North Dakota, South Dakota, California, Hawaii, and Oregon were absent from that first Task Force meeting of the third phase and one or more of these states were absent intermittently thereafter. After the governors had reendorsed the Task Force work on 1 March at a National Governors' Conference meeting in Washington, demanding more speed in implementing the new energy office, dissension mounted rapidly.

The Task Force aimed to have new WEPC staff on line in the spring and to be incorporated in June. From this point on, WEPC was known as the Western Governors' Policy Office (WESTPO). On 24 January, the Task Force decided to charge each state $12,000 dues for the first year. At the next meeting, Dale Mallicoat of Oregon voiced his concern about this cost associated with consolidation; he repeated, "We do not intend to subordinate the energy policy work in the Northwest to a broader Western states policy study." In contrast, David Freudenthal of Wyoming commended the Task Force and ISC for their haste in implementation and stressed the need to guard against WINB "driving" the proposed energy office.

In spite of a telegram from Governor Judge in mid March urging delegates to attend the critical meetings ahead, California, Nebraska, South Dakota, Idaho, North Dakota, Washington, and Hawaii were each absent from at least one. Judge suggested that the West hurry ahead to resolve dues allocations, legal aspects of consolidation, and matters of staffing a new energy office. He was anxious to proceed because, as Burgess noted, "Some federal officials and others are very interested in the status of the consolidated effort and concerned about *who* represents the Governor's regional energy organization at this point."[46]

According to the legal advice of Andrew Turi, Task Force counsel, David Engdahl, WINB counsel, and John Watson, former WGREPO staff member, gubernatorial action could not lawfully abolish WINB, but integration was possible through interlocking directorates within WESTPO. Noel Clark of Nevada and Dale Mallicoat of Oregon, however, worked against proposals to reform the existing board representation in WINB. It was apparent that Fred Gross, the former leader of WINB, was working to frustrate progress toward consolidation. Heated exchanges took place between Romer and Gross, and between Monaghan and Gross.[47] Eventually, with the tenacity of Monaghan and the support of seasoned regional WGREPO actors such as Metzger of North Dakota, Zee of New Mexico, Peters of Nebraska, Hall of Montana, Warnock of Arizona, Bucks of South Dakota, and Freudenthal, staff selection turned out in favor of the reformers. Representation patterns on the board of directors were altered also. It was not until after the September (1977) Governors' meeting in Anchorage that the Task Force's Energy Steering Committee met with WINB in Richland, Washington, and set substantive changes in the organization in motion. At the Richland meeting the name of the organization was changed to Western Interstate Energy Board/WINB, and Hawaii, Nevada, and North and South Dakota were formally asked to enact legislation recognizing the compact.[48] A major shift in the mission of the board of WINB was also instituted. In the future WIEB/WINB would be the West's major instrument for addressing federally related energy decisions.

Oregon's Mallicoat argued that original WINB states would incur costs of ten new staff members with the formation of WESTPO rather than a reduction of MSO staff. Disagreements over saving costs, consolidation efforts, new offices under the WESTPO umbrella Office of Policy Management, and the demise of the WGC became more strident in the spring. Governor Straub of Oregon sent a letter to the Task Force, which complained of unfulfilled promises and dissatisfaction with the elasticity concept in membership, defended the Northwest Energy Policy Study, and concluded that perhaps future decisions of the group "may restore our interest."[49] Nevada's Arkell registered formal complaints with staff proposals for a new natural resources office, challenged the need for the Touche-Ross Company auditing contract, wanted state dues reduced to $6,000 per year, and opposed staffing the new Office of Policy Management (OPM). Gordon of Alaska challenged Arkell's complaint, implying that the Nevada delegate had not briefed his governor sufficiently. Bill Peters of Nebraska and Tom Charlton of Wyoming were also somewhat concerned about the growth of a large central OPM staff but they favored proceeding. Colorado, Alaska, Utah, Arizona, Montana, New Mexico, and Idaho and North Dakota to a lesser extent were prepared to commit $12,000 dues. They considered the fee a bargain for the policy management advantages the West would receive.[50]

Maintaining momentum was essential to the reformers; the detractors wished to slow it down. The defenders of WINB, usually those who were skeptical of WESTPO's objectives, created serious administrative obstacles frequently through the spring and summer of 1977. Like Monaghan, Burgess and Chairman Romer incurred the wrath of officials in a number of states. Governors Hammond (Alaska) and Judge and former governor Rampton contributed to momentum, lending legitimacy to the Task Force at crucial points.[51] Governor Lamm was obviously excited by the progress and intentions of the states, but he was primarily occupied with leadership of a Drought Action Task Force at this time.

To legitimate their actions further, Task Force members who supported reform constantly referred to the mandate granted them in December 1976 by the governors. While they recognized that this continual reliance on the 10 December mandate was wearing thin by the summer, there was a necessity to use all means available to indicate progress before the governors met in Alaska. Lamm and several Western States' officials successfully completed a major multistate project on drought policy through the Drought Action Task Force (DATF) during the first months of the year, thus offering concrete evidence of their efforts. The success of this venture in federal-regional negotiation was timely, gave an example of the benefits of multistate policy management, and added momentum to Task Force work.

In a number of ways, the accomplishments of DATF were due to the same multistate organizational arrangements advocated by the WESTPO group. The success of the drought actions program offered insights on how the WESTPO umbrella concept would work. In fact, the key WESTPO office for water policy, the proposed Water Resources Policy Office, working with Governor Lamm and the staff of the proposed Office of Policy Management, conducted a model regional project on behalf of the West. In concert with the Carter administration, the DATF and Western supporters asked Congress for $844 million in loans and grants to help communities, farmers, ranchers, and businessmen stricken by drought in the Western and Plains states. In order to bolster Task Force momentum and to demonstrate the potential of WESTPO, Chairman Romer and Vice-Chairman Gordon highlighted the success of the drought project in May, urging members to bring the record to the attention of the Western governors. The response to the drought, they noted, demonstrated the value of the "accordion" approach to participation—where we have been able to include several States outside the Task Force (Minnesota, Illinois, Texas, Oklahoma, etc.);

demonstrated the value of the proposed Institute for Policy Research and gives substance to the idea of a "stand-by" research capacity—as indicated by the products of the comprehensive and timely legislative analysis that was initiated and is continuing;

demonstrated how the use of ad hoc groups—like the Western Regional Drought Action Task Force—within a comprehensive institutional arrangement or "umbrella" can help prevent the proliferation of permanent MSO's, staffs, budgets, etc. In this regard, note that the approach to the Drought Action Task Force has included in practice the functions we have proposed for the Western States Water Council, the Office of Policy Management, and the Institute for Policy Research within the larger WESTP.O framework;

demonstrated the need for institutions with the legal capacity to lobby and to achieve a delivery or implementation capacity in Washington—an ability to follow-up on State directives in Congress and the federal agencies;

demonstrated how State personnel can be mobilized on a temporary basis to help staff up a cooperative, multistate effort without adding permanent staff to our multistate organizations;

demonstrated how gubernatorial initiatives can be sustained through multistate staff level activities that remain accountable to State personnel—and particularly to the Governors and their representatives (in this case, the State drought coordinators); and

demonstrated the need for stand-by resources—both people and money. For example, more than a dozen highly qualified people and more than $35,000 have been mobilized for these efforts so far and their results have won our States considerable respect in the White House, in the federal agencies, and in Congress (especially in the Agriculture, Public Works, and Interior Committees).[52]

While Governor Lamm played a leading role in the drought

action project and both Judge and Rampton continued to provide visible momentum to the struggling Task Force, Governor Hammond and his aide Bill Gordon played increasingly more significant roles in the organization. Hammond had assumed the chairmanship of the Western Governors Conference after the Jackson Hole meeting and was to host the 1977 conference in Anchorage. Since his participation at Jackson Hole, he had enthusiastically approved the task force concept. Even though he was the only Republican governor in the sixteen-state group, he aligned himself comfortably with the Rocky Mountain contingent of Democrats and strongly endorsed the policy management concepts espoused by the Task Force. His chief aide in energy matters was Bill Gordon, chairman of WGC's Staff Advisory Committee and a leading member of the ISC, which was attempting to reform WINB. In his position as head of the WGC Committee, Gordon was a key figure in the proposed reorientation and reconstitution of the WGC. As a major supporter of WESTPO and in an effort to advance its progress, Gordon reduced the role of the Western Office of the Council of State Governments to that of "logistical supporter" of the Anchorage meeting. Gordon preferred to see conferences of governors "action-oriented" and supported by good staff analysis of Western policy issues—something which contrasted with the traditional WGC approach of a "forum of peers" and "huddling," as it had been termed by one aide.[53] With Gordon as head of the committee it was virtually impossible for the few supporters of the WGC to mount a strong defense against the WESTPO plans for its reconstitution and reform.

Nevertheless, June 1977 saw the major defensive actions of the WGC and the political setting for a tumultuous conference of the governors in Anchorage in the fall. In April, Jerry Norris, director of the Western Office of the CSG, the traditional organizer of Western Governors' Conferences, informed Bill Gordon that Governors Straub, Link, and Ray (Washington) had requested an analysis of *Regional Policy Management: The Task Ahead*, the report that concluded Phase II of Task Force work. In June, Norris mailed copies of the analysis to all the Task Force members, and staff advisory committee members. Norris reported, "We are . . . willing to address the issues presented in *The Task Ahead* to assure that the Governors of the Western

States *know the alternatives that are available to them* [emphasis added] and to assure that they receive the best possible service at the most reasonable cost."[54]

Norris implied that some states were suspicious that insufficient alternatives had been offered and that the Task Force was "railroading" the WESTPO concept of a new "umbrella" organization for the West. While recognizing the West's need for better controlling mechanisms for the region's MSOs, Norris argued that the existing Staff Advisory Committee of the WGC could do the job. Burgess's ideas, Norris intimated, were largely drawn from previous MSO reformers in the Southern Governors' Conference.[55] Burgess defended the rationale for WESTPO, pointing out reasons why the needs of Western states should differ from those in the South. Norris rejoined that Burgess and the Rocky Mountain group had given little consideration to using the WGC as their principal mechanism for MSO reforms and that the report submitted on 10 December 1976 contained numerous inconsistencies and confusing terminology.[56]

While it was true that Task Force staff reports were rather voluminous at times, there could be little argument with the quality of their analysis and the range of alternatives that were offered to the governors. Yes, it was plausible that the Staff Advisory Committee might have been reformed to offer more attractive services to the states. Most governors, however, were ambivalent about the WGC, and the few who demanded more research, staff analysis, and issue discussion were far more energetic and adventuresome in their efforts to be innovative than the silent majority. The committee had not offered an institutional response to the challenges facing the Rocky Mountain and Plains states since its inception in 1975 and had focused most of its meager efforts on conference planning. The WGREPO group, on the other hand, had accumulated extensive experience in multilateral problem solving and had gathered considerable momentum since 1974. This vigor had not been matched in Idaho, Oregon, or Washington; California and Hawaii were even further removed from the emergent threat of massive industrialization facing the energy-producing states. Idaho had chosen a unilateral approach to problem solving in both the WGREPO and WESTPO drives. There was no doubt that the significant differences in attitude toward problem solving between the states was working against both the WGC and WESTPO. But the Task Force of WESTPO was backed by a cadre of well-motivated states that stole the initiative away from the WGC and its director, Jerry Norris. To add to the fray, the continued dissent of Chris Smith of Idaho, Bud Kramer of Oregon, and Jason King of Washington on a number of Task Force initiatives did not augur well for the governors' conference in Alaska that climaxed the third phase of work.

The intragubernatorial tensions and suspicions prior to this Anchorage meeting were reminiscent of the creation of WGREPO. Lamm and Judge were regularly in newspaper headlines issuing scathing attacks on the Carter administration's stance on the development of the West's natural recources. Lamm complained, "Colorado and other Western states will not tolerate further water exports except under a clear and overriding national interest. . . . Western resource states are a sort of sacrificial area for the production of energy. . . ."[57]

Lamm was envied by some governors for his flair and his propensity for capturing extensive publicity and censored by others for speaking on behalf of his neighbors without their consultation. Some questioned his motives in backing WESTPO, and some observed that his presence in WESTPO reduced the inclination of certain Pacific Rim governors to participate.[58] Judge, who was a close supporter of Lamm's sentiments, took Governor Evans of Idaho to task for his withdrawal of support for WESTPO and for Idaho's accusations of the task force's untrustworthy performance. In a letter to Evans, Governor Judge sharply disagreed with Evan's views, deploring "some WINB members who refuse to accept that the new energy office, like WGREPO, will be an organization of Governors."[59] The governors of Washington, Oregon, and Idaho were said by some to have been offended by the claims of the Rocky Mountain governors that theirs was an "energy-producing" area whereas the Pacific Rim were "energy consumers." The Pacific Rim group reportedly felt that WGREPO had a "holier than thou" attitude.[60]

The Anchorage meeting of the governors occurred at the junction of several different streams of action. The traditional Western Governors' Conference had been a time for informal discussions of common interest to the Western states, a social gathering some called "huddling," and an opportunity for each governor to make a few newsworthy announcements that the press could publicize. Each governor would normally defer to his colleagues, thereby permitting every governor to promote an issue or resolution with news potential. Unfortunately, Task Force members did not appreciate this tradition. They were convinced the governors "should" have research data, policy analysis, issue identification, and alternatives established for implementation. This was how Lamm, Judge, and Scott Matheson of Utah liked to do business, and the Task Force tried to compile reports and policy resolutions from MSOs prior to Alaska for distribution and discussion at the WGC. The results were voluminous. The maze of reports, position papers, and resolutions alarmed several governors. The Task Force staff had, moreover, nominated governors to lead the discussion on separate issue topics as had been done effectively in the WGREPO. Some governors reacted by feeling "staff driven" and fearing they were losing control of their organization. Some objected to speaking on a topic that was not a personal specialty. This specialization interfered with the system of gubernatorial deference that had traditionally enhanced the progress of the meetings. The growth of paperwork looked to the governors like the creation of another bureaucratic organization when most of them were eager to divest themselves of the myriad MSOs that already existed. With the additional slip between the cup and lip brought about by cases of poor staff briefing of some governors before the meeting, the Anchorage meeting became a major turning point in the future of WESTPO.

The early highlights of the conference reflected the continual conflict and cooperation that persists between the Western states and the national government, particularly the U.S. Department of the Interior. At first, Secretary Cecil Andrus, formerly of the WGC when governor of Idaho, was warmly commended for his efforts to communicate with the states by Governors Lamm, Judge, George Ariyoshi (Hawaii), and Dixie Lee Ray (Washington). Lamm spoke effusively of the success of the Drought Action Task Force, the cooperation of the federal government and its $844 million appropriation, and the lessons learned that speak for a WESTPO-type regional organization: "See what the West can do when the Governors of states get together and go to meet the feds." Judge enthusiastically supported Lamm and spoke of the success of WGREPO: "We have seen the positive benefits of the pressure the West has brought to national energy policy. . . . see our impacts on federal policies and regulations to date." But then Herschler of Wyoming took issue with Andrus over recently released policy options for a new national water policy: "The policy options are dreadful, inadequate, inappropriate, a threat to a state's right to control water. . . . you want to replace prior appropriation with federal appropriation." In response, Andrus defended Interior's approach to the problem, then threw an angry challenge at the Western states and their institutional adequacy to solve problems of water, energy, and agriculture:

> Where is your [state's] constructive side? You states have not contributed to the design of national and Western water policy beyond emphasizing state's rights for dealing with consumption and irrigation. What are your comprehensive policies for energy, transportation, agriculture and water priorities? Look, most of the states in this room do not even have minimum stream flow laws. . . . to date your dialogue with us on national problems has been of little use. . . .[61]

Two and a half years earlier Rogers C. B. Morton, then secretary of the interior, had similarly challenged the states, and the establishment of WGREPO was the response.

While Governor Lamm encouraged the unity of the sixteen Task Force states for the sake of the West, he spoke guardedly. He was aware of the need to work toward a compromise. It was rumored throughout the Anchorage meeting that the West was grimly divided. Ray was the first to move against the WESTPO concept. She registered discontent at being delegated by the Task Force to speak to oil transportation issues when she preferred to lead a discussion of nuclear energy and waste disposal in the West. She attacked a stress on conservation in energy policy, argued in favor of the "technological fix" syndrome to society's problems and then launched a broadside attack at Governor Judge of Montana. Ray was critical of the neoisolationist stance that Montana had taken in the development of its natural resources: "I will subscribe to the sovereignty of a state, but not to isolationism. My state must import and export. So must Montana, Mr. Judge." Chairman Roy Romer of WESTPO spoke forcefully of what WESTPO could offer in enhancing gubernatorial control over MSOs and providing a mechanism for a united West. Ray interrupted:

> I am in doubt. I was not present at the December 10 meeting, and I am reserved about umbrella organizations. While I prefer to eliminate unnecessary MSOs, I will examine them myself. Superagencies like H.E.W. don't work. Do not create another devil here. The discussion of policy management is pretentious. I don't know what it means. I will not delegate authority to staff, and I want the Western Governors' Conference continued. Governors come and go like waves in the sea, but I am afraid WESTPO will go on, and on, and on. . . .

Governor Ray's critique hushed the large audience at the public meeting. Within a few seconds, Governor Straub of Oregon stood up. "I support Governor Ray. I am very disappointed by this inundation of paper at this meeting. If this sort of busy work continues like a super agency, I do not want to participate. We will not participate."

Romer tried to defend the WESTPO effort, reminding the governors of their earlier frustrations with uncontrollable MSOs. Judge corroborated this:

> Piles of paper and piles of organizations already exist. Now for the first time, Montana sees how much money and time we waste on the likes of WICHE and WASDA. I said no to a dues increase for WICHE and they went over to the legislators and got it there. I bet none of you can tell me what all these MSOs are doing. We have just got to coordinate policies for water, energy and agriculture. No more treating matters separately.

In reply to Judge and Romer and in response to Lamm's call for a compromise, O'Callaghan of Nevada interjected: "I will coordinate MSOs from my own state. Can't we accomplish all this without a coordinating staff [WESTPO]?"

Again, Romer defended his position, this time by retracing the rationale for the Task Force. Governor Scott Matheson suggested a compromise as Lamm had done earlier: "Regional organizations need to be thought through. We ought to serve our region well. Don't let's be at cross purposes today. Let's consider WESTPO on a trial basis and review it later."

With Herschler's concern about the mission of the Rocky Mountain Institute for Policy Research, Ariyoshi's complaint about the multitude of resolutions delivered to him by staff,[62] and O'Callaghan's desire to review the WESTPO budget, the governors adjourned to a closed luncheon meeting to continue discussions in private.

All reports confirm the tenseness of the meeting, complete with angry disagreement between some of the governors. Governors Ray, Straub, and O'Callaghan left the meeting early after declaring their intent not to participate in WESTPO. They expressed extreme displeasure in WESTPO's meddling with WINB and other MSOs. Governor Judge was impatient with the continual effort to accommodate the Pacific Rim States and pressed to reduce the membership of WESTPO to its original ten or eleven states. He also wanted the WESTPO supporters to withdraw from the WGC immediately. Lamm pressed steadily for approval of WESTPO, tried to soften Judge's impatience, and sought to make it possible for all sixteen states to negotiate an agreement. Eventually Ariyoshi proposed a compromise in which all states would agree to continue as members of the WGC for one more year and those who wanted WESTPO should participate in it. Known as the Ariyoshi Compromise, this plan gave both the WGC and WESTPO a lease on life for one year at which time WESTPO's performance would be examined. Governor Lamm announced this proposal at a packed, hushed, and news-hungry public meeting at which the sense of anticipation electrified the atmosphere. He tried to convey that the private session of the governors had been "courteous, reaching out, yet with strong feelings on both sides, in a community of interest for the West." For the first time it was clear that, officially, the managerial and organizational alignment of states necessary to control MSOs through sixteen states was not matched by the critical political alignment. From this point onwards, WESTPO was eleven states, as five of the sixteen declined to participate.

CONCLUSION

Not all the creators of WESTPO shared the objectives of regional unity to an equal extent. Once again Governor Lamm was a principal instigator, receiving strong suppoct from Rampton, Judge, and Hammond of Alaska. In contrast, O'Callaghan, a former WGREPO member, publicly worked against WESTPO's formation. As a result of the politics of creation of WESTPO, the West became divided between Pacific Rim and Rocky Mountain and Northern Plains subregions. The intricacies of conflict and cooperation between the sixteen states and the changing national and federal context were different indeed from the earlier WGREPO era.

Many Western governors and their key staff members testified to a lowering of anxiety as regards state relations with federal agencies during 1976 and thereafter. The WGREPO group felt headway had been made with negotiations in Washington, D.C.; the Carter administration's attempts to open decision making to state input exemplified such progress. The momentum to gather political unity for their states was urged on, in part, by the spectre of the frost belt phenomenon, the traditional desire for state autonomy, and the wish to meet federal authorities on equal terms. The evidence is overwhelming, however, that Western congressional delegations and members of key congressional committees considered the WGREPO and WESTPO actions of little consequence. There was little respect for the gubernatorial instruments as effective policy determinants for a region.

While the WESTPO era was marked by a slight easing of federal-state tensions, intraregional anxieties in the West came to the fore. At least two factors here draw attention to the geographical dimensions of the region exhibiting unity in the face of rapid economic development. First, Pacific Rim states were clearly not compatible with the Rocky Mountain group; Northern Plains states, while remaining with their WGREPO partners, were but mild supporters of the WESTPO arrangement. Second, the WESTPO Task Force deliberately proposed an elastic concept of opportunities for state membership. It was decided that a state ought to have the freedom to participate in any one of the five functional office programs, or any other issue that might arise, but not necessarily all issues all the time. This effort was directed at reducing the weaknesses engendered by "the least common denominator" alliance as exhibited by WGREPO. In addition, the concept of the elastic region provided for a noncontiguous state like Alaska, Texas, or Louisiana to participate when it so desired.

The elasticity of WESTPO eligibility was, in part, a timely recognition of the need to accommodate differences between Western states. The Task Force members had focused, not on forging identities of interest as had the WGREPO experiment, but on accommodating the common interests that existed while recognizing differences between states. The consensus was that WESTPO would handle "regional" problems. Energy development may have created a number of "common" or "shared" problems, but few, if any, required a regional solution. Consequently, the WESTPO era distinguished itself from the WGREPO era by a recognition of the need for multilateral actions, with less emphasis on the proper configuration of member states.

As was the case in the WGREPO era, states participated in WESTPO for a variety of self-interested reasons and with varying degrees of enthusiasm. Colorado, Utah, Montana, and Alaska were key states. New Mexico, Arizona, and Wyoming were also strong advocates of WESTPO. On more of a middle ground stood North Dakota, South Dakota, and Nebraska, their membership fees paid through the Old West Commission in the first year. Hawaii's participation fluctuated: although a paid member of WESTPO, it was split in loyalty to the Western Governors' Conference while Ariyoshi was chairman in 1977-78. If Nevada had become a participant and Hawaii had not, WESTPO would have consisted of all the "original" WGREPO group, plus Alaska.

The politics of creation of WESTPO were even more telling than those of the earlier institutional experiment. This was especially true in the areas of personnel. The Task Force and staff carried most of the momentum, occasionally gaining impetus from an encouraging work session with governors. WGREPO's effectiveness had been plagued by tension between Chairman Guy, placed there by governors in spite of protest, and the state alternates. From WGREPO's inception, the state alternates objected to Guy's participation. Guy had hinted his displeasure at the prospect of the newly emerging WESTPO, a staff-driven organization, and the demise of WGREPO. During WESTPO's evolution the staff played the principal role in personnel selection and the organizational design. The Task Force endorsed the strong leadership of Philip Burgess, largely because he held their trust and responded to their commands. In addition, the variety of operating styles between the governors was a significant determinant in WESTPO's evolution. Those preferring analysis, staff research, and follow-through pressed strongly for WESTPO's creation. On the other hand, some exuberant staffers failed to appreciate the traditional lure of the annual forum, the WGC, with its relaxed agenda and politics of deference.

The system of deference to colleagues was also a mark in the evolution of initiatives between the governors in the WESTPO era. Lamm, a dynamic and creative leader, had largely instigated the WGREPO and WESTPO activities and played a leading role in guiding them to fruition. During 1976–79 he was more of a mediator urging compromise and goodwill between his colleagues in neighboring states. His deference to the senior governor, Rampton of Utah, and his endeavor at compromise among the sixteen states at the Alaska conference, exemplified a maturity that developed during his gubernatorial experience.

6

The Institutional Response of a Region: An Interpretation

During the years 1972–79, the Western states passed through an extensive period of institutional reform, and their actions have been the subject of widespread debate and commentary. In this chapter we examine the views of opinion leaders in the West, and elsewhere as well, on this subject, and draw some concluding observations from their assessments. The objectives of the regional organizations and their promise and performance are taken into consideration. All of this inquiry, and our findings, raise important questions about the West and the notion of regionalism. The gubernatorial initiatives of the 1970s have demonstrated a resurgence of state participation in the federal system, but conclusions about regionalism and sectionalism are more difficult to draw. More detailed information on the evolution of regionalism and sectionalism in the United States can be found in the Appendix.

THE INTERVIEWS

For purposes of this chapter, approximately one hundred interviews were conducted. There were also discussions with most of the governors and their key aides who were assigned to the design of WGREPO and WESTPO. Interviews outside the gubernatorial and staff circles were conducted with representatives of newspapers, agricultural unions, labor organizations, the League of Women Voters, legislators, industrial executives, environmental groups, federal and state officials, and academics. The number of respondents was about evenly divided among the key states and included numerous federal officials as well. No claim is made for the statistical validity of the sample of respondents, but the respondents do represent a wide spectrum of interests. Respondents' statements were consistent with views and judgments frequently expressed to the author as he traveled the region. Rather than present individual responses, the chapter will specify predominant conclusions, identify significant differences in awareness of related events between states, and relate these to the broader theme of this study.

The interview schedule contained six major questions. The first inquired whether the respondent was aware of WGREPO or WESTPO, and if so what the perceived purposes of these governors' organizations were. The question was designed to reveal the extent of public knowledge of these organizations. Second, respondents were asked if they shared the view that the West, or their own state, had been subjected to discrimination, or benign neglect, at the hands of the federal government or industry. Was there widespread agreement that the region had been colonized? The third question probed for a sense of regional consciousness in Western state government and people. Had the history of their states, or the greater area, exhibited a multilateral tendency in problem solving? Fourth, to assess the support of Western legislatures for gubernatorial initiatives, respondents were asked if legislative appropriations might be forthcoming that would favor WGREPO or WESTPO. In consideration of the dominance of Democrats in governorships, the fifth question asked respondents to what extent, if any, fortuitous political circumstances may have influenced the creation of the two organizations. Finally, respondents were asked to identify the principal inducers and inhibitors of regionalism in the West. Which were more predominant, the inducers or the inhibitors?

WGREPO, WESTPO, and Their Objectives

One reason for the first question—on the knowledge one had of WGREPO and WESTPO—was to identify respondents who did not know of the organizations and their initiatives. Only four respondents were unaware of them—one each in South Dakota, Nebraska, Hawaii, and Nevada. There is, it would seem, widespread knowledge of the organizations throughout most of the states.

More important, and with major implications for the conclusions of this study, was that approximately seven out of ten respondents share the view that the governors' organization objectives were primarily motivated by the urge to increase state bargaining power vis-à-vis the federal government. In other words, center-periphery relations are perceived as a significant theme underlying the political actions of the West's governors. The overwhelming consensus was that the governors were trying to build up their power in an adversary political system, which accorded the sparsely populated Western states relatively low political influence in national affairs.

Respondents also felt that the governors' motivations were genuine in creating a forum through which to express regional interests and to make them known to national decision makers. Few respondents saw industrial corporations as the "common adversary" of the states; environmentalists in Montana and Colorado made up a large percentage of those who did. These same respondents expressed the view that energy corporations had practiced "playing one state off against the other" and that governors had united to avoid this practice.

Another set of responses included perspectives on the coordination and communication aspects of the two organizations. Most often, the need for uniformity in state regulations and administrative practices was seen as a probable reason for the existence of WGREPO and WESTPO. Another view shared by a number of respondents was that exporters and importers of energy in contiguous states would benefit from regional policies. In contrast, there was little awareness of the record of WGREPO or the linkages between it and WESTPO.

Although the topic may have been implied in the responses about the center-periphery balance of power, few respondents referred directly to states' rights. They tended to see the problem as multistate in nature. But few respondents offered the view that the governors were using WGREPO and WESTPO primarily as "soapboxes" from which to gain publicity rather than as purposeful service organizations for the states.

States in the West as Objects of Discrimination

The responses to this question varied considerably. There was no consensus that the Western region had been the object of discrimination by "the East," the federal government, or private corporations. Among respondents from the northern tier of states, in particular Montana, there was a strong feeling of federal and industrial heavy-handedness. The people of Montana and, to a lesser extent, Wyoming have been suspicious of big business for decades.

As regards discrimination against Western states, several themes emerged. The first is related to the extensive acreage of public lands in the states, except in Nebraska, Hawaii, and South Dakota. In general, each of the states continually seeks more control in decision making for the management of public lands within its borders. There were also subregional concerns. The strongest accusations of "benign neglect" and "exploitation" came from respondents in Montana, Wyoming, Colorado, and North Dakota—states which have been the center of coal and oil shale development proposals since 1970 and which stand to receive the most severe impacts of such development. Both federal and industrial forces were seen as exploiters in these four states. On the other hand, respondents in Nebraska, South Dakota, Arizona, Nevada, and New Mexico did not see their states as objects of exploitation. On the whole, there is strong desire within the region for more state control of decision making for use of public lands, a continual push in Nevada and Utah to have these lands assigned to the state, but no prevalent sense of corporate exploitation.

A second view on which there was consensus related to "the East's concept of life in the West." The claim was that the East (principally officeholders in the federal government and Congress) does not understand the needs, life-style, and priorities of the states in the West. Respondents were of the opinion that national decisions affecting the sparsely populated and rural

West, although mostly without intentional discrimination, seldom have been based on a full understanding of the region's preferences and needs. There was a widespread acknowledgment among respondents of the relatively low political influence the Western states have in Congress and of their weak position vis-à-vis the balance of the nation should they be called upon to extend rapidly the production of raw materials.

Plans to extend the production of energy since 1970 were seen by a number of respondents as more the result of the Nixon and Ford administrations than of traditional federal policy or "Eastern" pressures in general. These respondents believed that the political influence of the nation's energy industry reached unprecedented levels in these administrations. Respondents did not confirm a historical pattern of resource exploitation in the West, but they did argue that in the Nixon and Ford years pro-development objectives of federal land agencies had prevailed over conservation and wise management.

A few respondents were of the opinion that states in the West had received disproportionately favorable financial subsidization from federal sources over many decades. Statistics show that the region has been the beneficiary of more federal aid per capita than the national average and more aid than it has paid in taxes. These respondents argued that the area of public policy in focus was the crucial determinant in deciding the issue of discrimination. In areas of public policy such as rural finance, irrigation, and transportation, the West has been favored by Congress. Development generated by the demands for energy is the most extensive, and perhaps one of the few potentially exploitative drives the West has had to cope with in its history.

Legislative Support of WGREPO and WESTPO in the Future

More than half the respondents indicated that they would expect their state legislature, if requested, to supply funds to these organizations. The respondents who were of the opinion that their state legislature would probably not allocate funds were from different states throughout the West, but the only states in which they outnumbered positive responses were Nevada, Utah, and Hawaii.

A number of observations taken from these interviews provide useful background information on the reception these organizations have received and some key factors that may influence legislative response to their future. First, there was a widespread "wait and see" attitude toward the new organizations and a strong skepticism in expectations of their performance. Second, insights on the legislatures themselves reflect some important executive-legislative relations and contrasts in Western states. Most of the legislatures were seen by respondents to be conservative, even in states with liberal governors. According to respondents, there is a predominant development-orientation in the legislatures of most Rocky Mountain and Great Plains states. In addition, in the view of both respondents and scholars, state legislatures in the West are generally rustic and semiprofessional, and traditionally they have not expressed interest in appropriations for future-oriented projects involving surveillance, research, or planning. WGREPO and WESTPO had such a mandate; further, they were multistate mechanisms of the kind which Western legislatures have seldom favored. Respondents in Nevada, South Dakota, and Nebraska were of the view that energy development problems per se would not receive favorable hearings in their legislature but that threats to water supply might lead to legislative support for WESTPO.

Because there are basic differences among some potential producer states on attitudes toward extensive energy development, the range of positions WGREPO or WESTPO might adopt and the extent to which legislatures would tolerate such a policy are each limited. If the federal government stepped up pressure for increases in the rate and magnitude of fossil fuel exploitation, it is widely assumed that most legislatures in the West would support the governors' organizations financially. But if the organizations took a position too far off-center on any major policy issue, some state legislatures would most certainly withdraw their support.

Significance of Democratic Gubernatorial Dominance

With each participant governor in WGREPO or WESTPO through 1978 a Democrat except Hammond of Alaska, it was necessary to ascertain to what extent, if any, party politics led to the multistate organizations. Respondents were asked whether they saw this fact as a major determinant in the creation of either organization. Their answers were divided equally between yes and no, and there was no obvious correlation between the respondent's answer and his party affiliation and that of his governor.

Those who maintained that party affiliation made no difference were generally of the opinion that the seriousness of impacts was of major concern to most governors; as the federal government was a common adversary, both parties would see the problems similarly. This certainly proved to be the case with Hammond, the lone Republican until 1979. These respondents expected a mix of Democrats and Republicans in WESTPO in the future. In addition, as one observer argued, within the group of Democratic governors there was an extensive range of platforms. This range was bounded on the prodevelopment side by Rampton of Utah and on the environmental, quasi-isolationist side by Lamm and Judge. There exists no clear-cut philosophical difference in the Western states between Democrats and Republicans on environmental issues.

Respondents who argued that party similarity was necessary for the creation of WGREPO and WESTPO gave little supporting evidence for their observations. Some pointed out that with the extremes of platform, party affiliation was essential for a multistate organization. These respondents, in the main, surmised that WESTPO would disintegrate in the event that the states elected more Republican governors.

A very small percentage of the interviewees showed an in-depth understanding of the roles different governors played in creating WGREPO and WESTPO. Only one in ten had any knowledge of the leading roles played by Lamm and Judge. And an equally small percentage was aware of the conflict and compromise that had taken place between the governors during the evolution of the organizations.

Regionalism in the West: Inducers and Inhibitors

Respondents' perceptions of regionalism in the West could be placed into categories: 20 percent agreed without qualification that, yes, there was regionalism; 40 percent agreed with qualifications; 20 percent saw no regionalism and recognized starkly different subregions in the West. The remaining 20 percent of respondents had no views on the subject. While it is very difficult to separate the respondents into categories, it is helpful in understanding the findings on the characteristics of the main inducers and inhibitors to regional action, as seen by respondents.

Those who did perceive a regional consciousness in the West represented both northern and southern tier states. Respondents in Arizona and New Mexico claimed a noticeable drift toward interstate cooperation in the Southwest and a regional consciousness developing in the people and their leaders in recent years. This evolving regional consciousness was the result, in part, of the realization that many problems of public affairs were parallel across state borders. It was also due to the realization that only through unified approaches to national decision makers would the sparsely populated Western states gain political stature. The strong cultural traits of individualism in the West had contributed to the delay of regionalism before the 1970s, but the decade now saw, as respondents in Wyoming, Montana, and North Dakota testified, a rapidly developing regional consciousness, especially in the northern tier states. The increasing regional actions by environmental groups and labor organizations and the catalytic role played by NGPRP were offered as evidence of this new regional consciousness in both the people and their leaders.

Those respondents who acknowledged a qualified regionalism —the largest percentage—had numerous comments. Most reported on indications that the people and government leaders had come to recognize that the West faces common problems as a result of energy development, but they claimed that until 1974 states had largely pursued their interests and dealt with their problems as individual entities. They could see signs of a regional consciousness but could not agree on one principal cause. Among the causes named, dislike of the threat of federal preemption, bargaining for federal research and development funds, preference for orderly mineral development, and the existence of a crisis were the most often emphasized. The crisis, perhaps the item closest to a principal cause, was seen as the threat of federal preemption that would force states to carry out development at unreasonable levels.

A related theme that emerged was the perceived growing intensity of federal industrial pressure. The federal government was viewed in many states as a common adversary to the West—a sense which has stimulated notions of regionalism. Whereas certain respondents attested to the growth of a regional consciousness—in effect, a regional defense against the nation's power centers—they commonly expected that unity between the states would disintegrate once the crisis abated. A large proportion of respondents expected that, once the common adversary was gone or problems between the region and the nation were resolved, the governors and administrations of various states would begin to squabble over resource allocation problems.

While there is growing recognition of some indicators of a regional consciousness—and some implicit agreement that energy development is the catalyst—few respondents foresee a permanent ten-state regionalism. A small proportion of the interviewees identified subregional groupings within the ten-state area, which, over a sustained time, would make for unlikely regional partners. The coal-producing states in the northern tier were seen as the most likely to maintain a multistate consciousness. Another division of the ten-state area was made by one respondent on the bases of the Great Divide and agriculture. In this cultural classification of subregions, Colorado, Montana, North Dakota, South Dakota, and Wyoming were defined as agrarian social systems linked to the rural states of Nebraska and Iowa. To the west of the Great Divide, Utah and Nevada, which also contained rural and agrarian cultures, were strongly influenced by irrigation and reclamation and lived by a land ethic different from that which prevailed in the states east of the Great Divide and along the front range of the Rockies.

The few respondents who perceived no regional consciousness in the West argued that state administrations had not yet collaborated purposefully at all. These respondents did agree that there is a widespread apprehension in the West owing to anticipated political pressures in the East, but they saw more of a regional awareness in Texas and its contiguous states than was exhibited in the West. They claimed that unanimity of thought on the part of Rocky Mountain and Northern Plains states was prevented by the competition for water and several traditional inhibitors.

In examining the views on regionalism, it becomes apparent that there are dozens of inhibitors and inducers of the phenomenon. The respondents have helped identify the most recognizable and significant of these factors. No consensus exists as to whether the inhibitors or the inducers are more prevalent currently, nor on which of them is likely to be dominant in the future. There is a strong consensus, though, that the inhibitors have been more dominant than the inducers in the history of the West to date. Respondents tend to expect inhibitors to dominate again in the future. It would appear that the 1970s were a period of transition in which the inducers came to the fore primarily because of the presence of an external threat.

The only inducer stressed by a major proportion of the respondents was the threat from the federal government that would force states to cope with unmanageable rates and magnitudes of development on public lands within their borders. Linked to this threat was the evidence that the Nixon and Ford administrations had actively supported preemptory legislation in areas such as air regulations and power siting. In other words, respondents generally view the West as the periphery united to defend itself from the center. Other inducers less often identified in the survey included the common culture of Western states as it is reflected in their strong attachment to a rural lifestyle, open space, and especially the dislike of "people pollution" or crowded living. States' rights, although traditionally defended in the West, appeared even more attractive in the wake of the loss of public confidence in the national government in the 1970s. Another inducement to unify the Western states' political power has arisen as knowledge in the West has spread of the potential pressure of national opinon that may be directed at outvoting Western congressional delegations.

The major inhibitors to regional consciousness and regional action, as seen by the respondents, are not new. Mostly they are traditional, philosophical, psychological, and parochial. Most prominent of the responses given was the view that state jealousies of rights and discretionary powers would always inhibit multistate collaboration. In addition, there are well-recognized differences in priorities between the states, and many respondents argued that water allocation problems would probably always halt unanimity of thought in the West. Once again, part of the parochiality of the states was blamed on the rustic and rudimentary characteristics of their legislatures.

THE ROLE OF WGREPO

The set of opinions and observations expressed above places the more particular performances and activities of WGREPO and WESTPO in context. As noted above, there was considerable skepticism in the West itself, and in the East, about the potential of these regional organizations. After two years of service, WGREPO's role was absorbed within the WESTPO arrangement, but during its reign its performance was observed by numerous key state energy officials, governors, federal agencies, and congressional committee staffs. What was WGREPO's contribution?

WGREPO's initial policy resolutions pointed to an urge to resist and oppose federal intrusion in states' rights, to establish the primacy of state strip-mining and environmental laws on federal lands, to become involved in national decision making, and to establish multistate coordination in dealing with regional problems that crossed state lines. Toward the fulfillment of these objectives, WGREPO was assigned numerous functions such as gathering data, operating as the nerve center in intergovernmental relations, and drafting policy resolutions which states could use voluntarily as guidelines for parallel laws. Even with a significant amount of multilateral information sharing that took place in its two-year operation, WGREPO was viewed in many different ways. Some Westerners regarded it as a focus on energy policy activity. Most observers, however, saw it as only one of the spokesmen for the West, and virtually no one agreed that it was the sole spokesman. Most governors were reasonably pleased with WGREPO's performance, although few relied on it as the spearhead of policy for their states. Most congressional figures considered WGREPO to be of little consequence in national legislative processes. On the other hand, most federal agency personnel felt that WGREPO did provide a means for them to deal with the states of the region, even though it was almost always necessary to deal directly with each state as well.

While it cannot be claimed the WGREPO brought cohesion to Western state efforts to influence national legislation on energy matters—and it certainly generated little in the way of parallel legislation or management practices between its states—the WGREPO apparently played a useful role in focusing and formulating positions on operating procedures for certain federal agencies. WGREPO contributed to the states' gaining considerable control of environmental protection of federal lands, and had input in the federal land development process. WGREPO's advocacy, efforts at negotiation, and continual monitoring of federal legislation and agency rulings contributed to state gains in at least four other major areas. The Energy Minerals Allocation Recommendation System (EMARS), an effort to streamline and improve federal coal leasing practices, was a constant object of surveillance by WGREPO as the system was developed and implemented. The Federal Coal Leasing Amendments Act, passed over President Ford's veto in 1976, was a piece of legislation monitored and supported by WGREPO that enhanced the interests of the West. Among other things, the act would increase the state's share of coal royalty payments from 37 ½ percent to 50 percent and give the states greater freedom in the expenditure of the funds. This change was typical of the results of WGREPO's constant attentiveness on behalf of Western governors; it gave the states more front-end capital for use in facility siting regulation, in impacted communities, and in the enforcement of land reclamation and rehabilitation.

Another success for the states, largely attributed to the research and advocacy of WGREPO, was the inclusion of Section 211.75 into Title XXX of the Code of Federal Regulations. At issue was the right of states to have preemptory power over land developments on federal lands. Proposed federal regulations were initially released in May 1976, and following a half year of determined state-federal negotiations, the federal authorities offered the states Cooperative Agreements late in 1976. In the agreements, state standards for the management of federal mined lands would be employed where they were more stringent than federal standards. The threat of federal intervention was further eased by the grant of procedural prerogatives to the states. Federal authorities granted that states could provide the inspectors, cite the violations, and approve the mining plans. By

June 1977, Wyoming and Montana had ratified agreements with the federal government that permitted enforcement of the more stringent state regulations. Colorado, Utah, New Mexico, and North Dakota had agreed to enforce federal standards because their state laws were less stringent than the federal regulations.

As another major contribution to state autonomy, WGREPO spearheaded a successful effort to reorient a section of the statute known as the Surface Mining and Control Act. Signed by President Carter in August 1977 after Ford had vetoed an original version twice, the law proposed to preempt all state and local control of environmental protection in coal mining. WGREPO, on behalf of the Western states, mounted a campaign of negotiations that climaxed with the insertion of a "grandfather" section in the statute. The states were thus able to demand that the Department of the Interior accept federally approved state programs for the control of coal mining of state, private, and federal lands. It was the responsibility of the states to submit their regulatory programs to federal authorities for approval by August 1979. No doubt the newly established Western Interstate Energy Board (officially referred to as WIEB/WINB) will coordinate this major undertaking for the states in the region.

Successes in altering federal agency procedures as outlined above were the most tangible indicators of WGREPO's role. Perhaps the regional posturing that WGREPO maintained had less tangible results, but it definitely contributed to placing Western issues in focus. As Director William L. Guy claimed frequently, "I don't believe the federal government has any [other] option than to work closely with states, because the states are in a position in so many ways to thwart federal programs."[1] As a spokesman for the Western Governors, although with limited credibility in the eyes of many congressional figures and the states' alternates to WGREPO, Guy insisted that the states would fight energy projects, if necessary, with the use of crippling air quality laws, water use restrictions, land use planning, extraction taxes, and environmental lawsuits. There is evidence to suggest the WGREPO's posturing, allied with the stands of individual governors, did move federal authorities gradually to accommodate Western positions.[2] It is fair to claim that WGREPO assisted the states to advance in two years from the position of weak critics to an emerging force that began to sharpen the weapons it had at hand to thwart unacceptable development. Rhetoric and posturing, some useful analysis of legislation, changes in federal regulations, and a role as a forum were not sufficient, however, to make WGREPO the ultimate political alliance or a regional institution commanding respect in national policy making.

Fundamental state differences over some of the finer details of growth and environmental issues and the pattern of predominantly unilateral behavior on behalf of Western governors inhibited WGREPO from becoming the locus of regional energy policy. Conflict occurred in the cases of New Mexico's electricity tax, which discriminated against Arizona, and the lawsuit between South Dakota and Wyoming over water use. Competition for synthetic fuel programs and research funds for solar energy and similar programs are still evident among a number of states. Economic benefit has been a major concern in Utah's policy, and at times this clashed with the environmental priorities and cautious progress on oil shale projects that marked the ecologically sensitive policies of neighboring Colorado. Montana and Colorado could endorse a "strip and ship" philosophy on coal development but Utah and its Senator Frank Moss could not. Many federal officials and congressional staff aides in Washington, D.C., testified to schizophrenic behavior by the leadership of Colorado and other Western states who wanted the financial, economic, and employment benefits of development but who undermined rational policy making by fiercely working against projects such as synthetic fuel commercialization.

Not only was WGREPO handicapped by the "lowest common denominator" phenomenon when seeking consensus on major policy resolutions, but it was also seriously fractured by internal bickering and dissension. This was not readily attributable to relations between the governors even though they each conducted the majority of their state-federal negotiations directly, bypassing WGREPO. Such was the case when Governor Lamm would speak for all or any one of the WGREPO governors, his own state, or as chairman of the NGC subcommittee on energy. A primary cause of the fracturing of WGREPO was the tension between state alternates and Director William Guy, a direct outcome of the original politics of creation of WGREPO. The alternates, generally young and aggressive, expected the low-key, formal Guy, formerly three-term governor of North Dakota, to answer to them. In contrast, Guy saw his peers, the other governors, as his employer. Where the alternates wanted to pursue a number of challenges and directions simultaneously, Guy chose to move more cautiously and incrementally. The anxieties that developed became apparent in some WGREPO alternates' efforts to undermine Guy publicly and in the efforts of a number of states to press their intergovernmental priorities directly to the federal government, thus bypassing Guy. The purse strings for WGREPO, held by the U.S. Department of Commerce, also limited the role Guy and the alternates could play. It was this fractured, out-of-focus image that WGREPO presented in most of its business undertakings. With its limited budget and the nonsupport from the inactive Western Congressional Caucus, the WGREPO, despite its modest expertise, was destined to become less than a powerful regional instrument.

THE PURPOSE OF WESTPO

Given the brief analysis of WGREPO's contributions to the needs of the Western governors, what is it about WESTPO that the states hoped would improve their managerial and political capacities? Which key factors contributed to the timing of the transition from one regional organization to the other? Did the performance of WGREPO, and its image, influence the design of WESTPO?

There is no doubt the WGREPO experience influenced the work of the Task Force. That a regional organization needed direct and frequent involvement of its governors had become obvious. The Task Force acknowledged this in its first study: "Effective regional organization depends on how Congress or federal agencies perceive the organization and the Governors' support for it."[3] Learning from WGREPO again, the governors were determined to divest themselves of dependence on federal finance for the core functions of WESTPO. The states had learned the necessity of having one locus for their policy decisions, thereby avoiding the difficulties WGREPO experienced as merely one of many Western organizations dealing with regional energy problems. Avoiding these difficulties required managerial controls of regional organizations in more than just energy; the cross-cutting problems of energy, environment, and economy demanded a synthesis in policy formation. Finally, WGREPO's lack of depth in analytical capacity taught WESTPO the importance of harnessing the scattered human and technical resources of the region.

Consequently, the states embarked on a more sophisticated managerial framework in the WESTPO era. They hoped to improve on the good points of each of their former regional experiments such as the Federation, WGREPO, the Western Governors' Conference, and key MSOs. The convergence of the conceptual and managerial skills of Philip Burgess, the trial and error of the Task Force, the evolving enthusiasm and momentum of key staff, and the special drive of Governors Lamm, Judge, and Rampton sparked the principal chain of events. The core personnel revolved around former Federation governors and staff, including former Governor Jack Campbell of New Mexico and, of course, the WGREPO group. They proposed that WESTPO should resemble an inverted funnel through which the region's MSOs would pass expertise upward for conversion into gubernatorial policy. Even though the level of anxiety between the states and the federal government had diminished from its levels in early WGREPO days, the perception of a threat from the East was stimulated by the growth of unity in the Northeast. Accordingly, Western states moved their response to rapid energy development onward from a single-issue focus to encompass a drive for more capacity for "policy management."

To appreciate the role and scope of WESTPO, one must understand its development between the Anchorage meeting in fall 1977 and its first board of directors meeting in Washington, D.C., in spring 1978. Even after almost two years of work, the gubernatorial conflict in Alaska had indelible implications for WESTPO's form and functions. The original form of WESTPO, which was proposed prior to Alaska, assumed a political alignment of sixteen states and an appropriate management structure. The eleven-to-five political realignment at Anchorage was not so readily matchable by managerial alignment. There were several key MSOs that straddled the line between the eleven proponents of WESTPO and the five who declined membership. If the Anchorage meeting had resulted in a unanimous consensus of states, WESTPO would have had the mandate to press ahead with standardized management practices, efforts at "sunset" review, and the like. But because of the political realignment, there were some major managerial consequences. Not until the governors' first meeting early in 1978 was the earlier enthusiasm of the Task Force era restored in the minds of key staff. Gubernatorial commitment, interest, and attention were necessary to accomplish the sense of vitality that exists in WESTPO today.

WESTPO's form and functions are purposely fluid, but fundamental patterns have emerged. Chartered in the State of Colorado as a nonprofit corporation of voluntary state members, WESTPO is best described as a unique general purpose organization. Governors constitute its board of directors, and their designees constitute the Regional Coordinating Council. Many council members are from the experienced Task Force group, and a large percentage of them were formerly principals in WGREPO. WESTPO is "driven" by "lead" governors and coordinated by the Board of Directors and the Council. The lead governors, as depicted in figure 3, have responsibilities for identifying policy and program priorities and for speaking on behalf of the governors. The small, flexible staff, formerly designated as the Office of Policy Management, have charge of producing studies as required, undertaking legislative analysis, convening seminars and conferences, and pursuing a "sunset" agenda. This set of tasks constitutes much of what was formerly described as "policy management." A deliberate decision to deemphasize the tone of "policy management" and to pursue the objectives more gradually and informally reflects the impact of the conflict in Alaska, as well as a number of lessons that the Task Force learned through its evolution. WESTPO uses the talent and services of MSOs and supplements and stimulates existing regional resources. Principal lead agencies for the subject areas, which act as reservoirs of technical talent, include the Western States Water Council, Western Federation of Human Services, and the WIEB/WINB combination (see fig. 3).

WESTPO was originally visualized as a gubernatorial organization with management responsibilities for MSOs. Since its inception, WESTPO has deemphasized managerial matters. It draws on the resouces of existing MSOs, while gradually developing the capacity to bring about consolidations and streamlining. The members of the WESTPO board of directors and council are al-

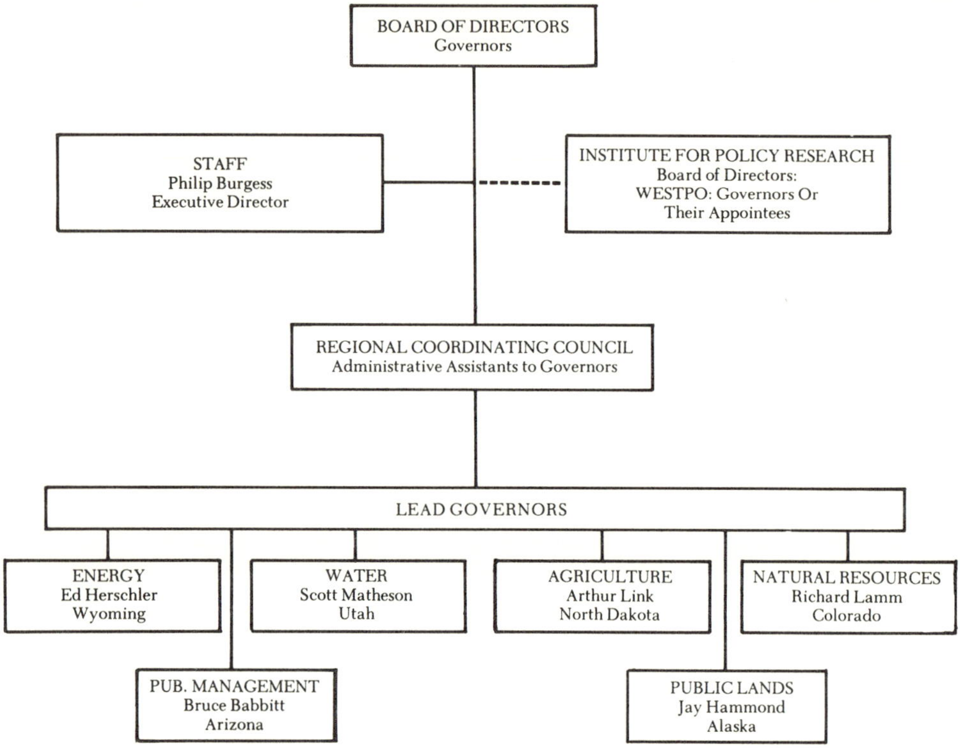

Technical staff support provided by:

- WIEB/WINB
- Western States' Water Council
- Western Federation of Human Services

Figure 3. Organization of the Western Governors' Policy Office, as of June 1979.

most unanimous in their conviction that their configuration of states is philosophically compatible. They hope to accomplish through common ideology, and commonalty of interest, what they discovered was nearly impossible through rigid managerial techniques. Given the crescendo of enthusiasm that followed the first meeting of the board at the time of the White House Conference on Balanced Growth in February 1978, it is evident that WESTPO has succeeded through informal procedures especially because of the close gubernatorial support the organization enjoys.

The White House conference was a leading indicator of WESTPO's role and potential. In setting up a forum for the governors, and as spokesman for the West, the new organization successfully demonstrated the utility of its role as an "umbrella" and orchestrator of the region's MSOs. The WESTPO report *Balanced Growth and Economic Development: A Western White Paper* was a synthesis of cross-cutting policy priorities. It was drafted to compile the talents and resources of the West's MSOs and state and federal agencies.[4] As was customary, some MSOs, in particular WICHE, "bitched all the way to the railroad," as they say in the West. It contributed reluctantly and exemplified the challenge WESTPO faced in its attempts to improve the accountability of renegade fiefdoms that resisted gubernatorial surveillance. The board could see firsthand the benefits of good staff work in WESTPO and the advantages of orchestrating and focusing the attention of MSOs on policy problems. Technical capacities in the West's MSOs have remained reasonably stable, while the priority of policy issues fluctuates and governors and their staffs come and go. In order to sustain a strong capacity for policy leadership, the governors demanded a high-level general-purpose mechanism that could cope with constant changes in issues and state membership, and fragmented MSOs. These themes were repeated in the governors' enthusiastic commendations of WESTPO at the White House conference.

In the early months of WESTPO's operation, the governors insisted that "sunset" procedures should continue gradually and showed their annoyance at some MSOs being like a "tail wagging the dog" (Governor Judge's term). They were not, however, interested in a rigid managerial control mechanism. The routine business of bureaucratic guild systems was not of interest to most governors, but that they be able to call on MSOs for a prompt and efficient response in times of need was essential. WESTPO provided such a linkage mechanism, facilitating prompt accountability. The first board meeting demonstrated this. The staff of WESTPO also found the pace of reform much slower than they had anticipated. Governor Judge pressed for complete disaffiliation from the Western Governors' Conference, but Governor Herschler objected. Apparently, the tremendous traditional attraction to the WGC had been underestimated by the WESTPO staff. The system of deference that influenced intergubernatorial decision making generally prevented a governor from moving too quickly in any direction in case it should be viewed as unreasonable by his peers.

Besides the enthusiasm that accompanied the performance of WESTPO at the White House Conference on Balanced Growth, there were three special indicators of its acceptance by its board of directors and of its potential as *the* spokesman for the region. First, it became clear that WESTPO need not be viewed as a threat to MSOs. In fact, cooperative, productive MSOs could expect the attention and support of governors in instances where WESTPO's coordinating requests were followed. Through WESTPO, governors were accessible to MSOs, thus giving organizations visibility and credibility. Second, the governors chose to lodge a new Regional Energy Advisory Board, authorized by the U.S. Department of Energy, in WESTPO. The intent of federal authorities was to enable Western states to contribute to national policies affecting their region. The new instrument had a policy-making function, so the governors gave the responsibility to WESTPO rather than to the WIEB/WINB or any other technically oriented MSO in the region. Finally, the joint support of the governors so apparent at their first board meeting was the most significant indicator of the organization's legitimacy and superior standing. The commitment, attention, and interest of governors gave WESTPO enormous potential. Gubernatorial attention was a magic ingredient that induced the participation of all manner of Western organizations and interests. This commitment of the governors was also a key indicator to federal agencies that the West was serious in its intent; commitment was the means by which the nation at large could judge the legitimacy and power of WESTPO.

In reviewing the institutional response of the West, first WGREPO and then WESTPO, two continuous phases of evolution emerged as prominent. A third phase—to maintain the *political vigilance* of the West, to increase the *policy management capacity* of the states, and to arrive at *one voice* for the region—will be necessary. Such an effort will take time, and the process will be vulnerable to interruptions, but the WESTPO innovation has the form and momentum provided by its board, council, and staff to launch this third phase. The ingredients for the prolonged success of WESTPO are shared with several other sources of influence within the region and outside it.

During the WGREPO era, the governors groped for an institutional means to implement a consensus. The focus of this consensus was on the improvement of the capacity to make energy policy, the defense of their states, and the contribution to national decision making. The second phase of progress, WESTPO's era, was much more sophisticated as it sought greater effectiveness through the methods described previously. While energy development still provided the focus, the need to cope with interrelated policy issues of natural resources, agriculture, and water was obvious. Beyond the substantive issues, the governors needed a management framework that effectively utilized the state's technical and scarce human resources and the participation of the private sector. WESTPO's early performance would seem to indicate that it will be a success, but there are a number of crucial inhibitors to the implementation of the larger managerial objectives.

When the political alignment of the sixteen states broke up, the feasibility of the preliminary managerial alignment was reduced. MSO participation remained problematic. The Pacific Rim and Rocky Mountain states appeared incompatible on numerous grounds, including contrasting views on energy development policies—a fact that would hamper MSO conformity. Equally incompatible were the governors' views on staff roles, policy analysis, and multilateral problem solving. The Pacific Rim group was not happy with strong roles and initiatives by staff, whereas the WGREPO group had demanded such, in their long-standing efforts to compete with the technical proficiency of federal agencies and congressional committees. Bureaucratic resistance on behalf of numerous MSOs, in particular WICHE and

WINB, frustrated progress. Divided loyalty to the WGC was a cleavage that Governor O'Callaghan and others used to slow up Task Force efforts in 1977, posing uncertainties for WESTPO during 1978. Equally important were other intricate aspects of the politics of creation surrounding WESTPO's evolution, in particular the role of deference by one governor to another, contrasting personalities, and the general preference of governors not to do battle with peers and colleagues. During 1978, in WESTPO's early reign, there were still too many voices speaking on behalf of the West. Even after some four years of work to focus gubernatorial attention in one institution, most governors resisted the struggles it would take to abolish duplicative MSOs.

To build upon the objectives that WGREPO and WESTPO have pursued, continual managerial and operational refinement will be necessary. As a part of this, sound and productive linkages will need to be established between WESTPO and the region's Title V commissions. The role of the WGC will need to be resolved. As it now stands, the West as a political bloc is not fully established, while the Western Congressional Caucus remains inactive. Nine of the governors of the original sixteen WESTPO states faced the electorate or were constitutionally restricted from further office late in 1978. A year later the new Republican and Democratic governors elected in WESTPO states indicated encouraging support for the organization. No partisan problems had arisen, and while new governors could be expected to institute some changes, WESTPO was stable and active. New alignments with the Pacific Rim states are not likely, although they should not be ruled out. If the politics of regionalism spreads as a national phenomenon, the West, represented by WESTPO, may develop a stronger cohesion and sense of common cause.

In order to maintain political vigilance in the Rocky Mountain and Northern Plains states, to ensure increased state capacity in policy management, and to arrive at a coherent voice for the region, WESTPO faces challenges that appear to be surmountable. First, it is imperative that WESTPO improve on the image that WGREPO carried in Washington, D.C. To do so, the governors need to demonstrate consistently their interest in and attention to the organization. This may be possible as WESTPO has been a genuine service to individual state needs; it is not as concerned with developing regional policy resolutions as was its predecessor. Should the organization be able to capture the attention of governors for MSOs, and in turn provide accountability and synthesis for the governors, it could establish a niche and constituency. Its capacity to deliver for the West will depend largely on its success in harnessing the technical capacity inherent in the region's MSOs, industry, and universities. WESTPO will need also to increase its visibility within the region and in the nation at large. Success in these basic areas would give WESTPO an opportunity to serve its states and region more effectively than any of its institutional predecessors. It would set a constructive precedent as a mechanism advancing the role of the states in the federal system.

STATE RESPONSES TO DEVELOPMENT: A HINT OF REGIONALISM

Indications of a regional or sectional posture in the Western states would include positions, resolutions, or policies on which there was a regionally distinctive consensus. There are certain issues, and some formal policy resolutions, through which WGREPO and WESTPO have demonstrated a gubernatorial consensus. Even though many of the energy-related problems in the states are held in common, or are shared, few require—or under the Constitution could obtain—a regional approach for resolution. Let us examine an obvious set of uniform responses by the states, and then try to uncover indicators of a political, organizational, and philosophical nature that relate to regionalism.

There are firm indications that all the states in the WGREPO/WESTPO group seek energy-related information. The producer states seek guidelines on the rate of development and its temporal aspects. The consumer states want to maintain open communication channels with the producers to ensure supplies. Both producing and consuming states require information on the primary and secondary impacts of development. They seek help in planning to mitigate problems and in funding to alleviate stress in boomtown areas. Information also is of crucial concern to each state for reasons of self-protection and the pursuit of its interests.

Within the region there has been a common response to the lack of federal energy policy. This response has been expressed in news releases, hearings, press conferences, and through both WGREPO and WESTPO. The governors as individuals, and now as a regional bloc, urge speedy information for national guidelines for energy management. The general demand has been made to have state input in federal policy formation in the critical areas of energy, water, economic development, agriculture, and natural resources. The states as individual entities and as an organized group have sought leverage and access with respect to federal decision making in each of these problem areas. States' rights to direct the future of energy development within their borders, and to have more influence on the management of public lands, have also been defended uniformly in the West. The participation in WGREPO and WESTPO and the negotiation over resolutions and research tasks indicate a measure of multistate consensus in these matters. The mere formation of WGREPO and WESTPO is also evidence of consensus on the difficult position of individual states with regard to keeping abreast of the action and inaction of numerous federal agencies on energy policy. The establishment of an information-gathering nerve center was therefore necessary.

Through their unified activities, Western governors have issued a common response to the need for extending state executive branch capacity. Especially noticeable in the WESTPO era, the governors sought new managerial tools through which to prod the responsiveness and accountability of their existing MSOs. They share not only the MSOs but also the burdens of their uncoordinated policy actions. This commonality of concern and the subsequent creation of the WESTPO policy umbrella stretching over state borders are unprecedented in Western politics and history. Through the National Governors' Association (formerly NGC) most of the states in the Union have examined executive branch affiliations in recent years, but no contiguous group of states can match Western initiatives for institutional reform.

The political, organizational, and philosophical dimensions of regional consensus are not plentiful. The rebirth of the Western Conference and the reassertion of representatives of Western Congressional Delegations are the only regionwide political actions to date. The election of Democratic governors is not a valid indicator of multistate party response as the wide range of priorities among them crosses party lines. The political difference between Rampton of Utah and Judge of Montana is one such example.

Some organizations have helped to promote a regional consciousness. Whether such organizations as the Federation of Rocky Mountain States, Old West, and the NGPRP were vehicles for the expression of a growing multistate awareness and concern, or were principal stimulants themselves, is uncertain. Whatever the interpretation, evidence suggests that these mechanisms did facilitate multistate cooperation although not uniformly between all of the states in WESTPO. No pattern of parallel legislation or creation of similar energy-related administrative arrangements within all these states exists. There is no common philosophical orientation toward energy development. Whereas Montana and Colorado have had quasi-isolationist and environmental protection priorities, Wyoming and North Dakota have had a cautious and more cooperative stance with emphasis on agricultural preservation. In contrast, Utah's view has been predominantly one of welcoming industrial growth.

The question of major interest—given these matters on which there is some consensus and given the political, organizational, and philosophical dimensions of state response—is whether the West is involved in regionalism, sectionalism, or neither. There is no one best definition of the distinguishing characteristics of regionalism and sectionalism. Let us test the evidence revealed by this study against the definitions of these two terms put forward elsewhere. We will examine the ideas of four observers of regional phenomena—Fulmer Mood,[5] Rupert Vance,[6] Donald Davidson,[7] and Frederick Jackson Turner[8]—and then the differences between sectionalism and regionalism discussed by Howard Odum and Harry Moore.[9]

Speaking in the House of Representatives in 1806, Josiah Quincy identified some of the key attributes of sectional sentiment:

> Whenever any great section of the Union shall deem itself neglected, and the opinion becomes general among the people that they are sacrificed or abandoned; that they have not any, or their just weight, in the national scale, a series

> of struggles must commence which will terminate either in redress or in convulsions.[10]

Our observations here have not revealed sufficient evidence to suggest that a majority of the people of the West feel they will be sacrificed. But neither do they believe that they have their just weight in the national scale. More likely, support is widespread in Colorado, Wyoming, and Montana for the view that energy-bearing states such as theirs would become sacrificial objects if they were not vigilant and vocal. The formation of WGREPO and WESTPO is a clear sign that participating governors seek more weight in national decision making. In this latter respect, energy development has acted as a catalyst of sectional behavior on the part of the governors.

Through his notion of maturation through history, Vance claimed that each major area of the United States passes from the frontier stage and becomes a section and ultimately a region.[11] There is no doubt that the states of the West have had a marked frontier era in political, social, and economic terms. If bloc voting in Congress is used as an index, however, there is no evidence that the West has ever been a political section. There is little to indicate that the West is about to step from one of these three stages of history into another, or even which stage it presently occupies. If the Western states are to vote as a bloc in Congress on matters pertaining to their interest, Davidson would refer to such action as political sectionalism.[12]

What Frederick Jackson Turner would define as sectionalism has not occurred in the West as yet, but some of his criteria are recognizable in Montana. They may become more widespread if the gubernatorial initiatives are followed by bloc voting in Congress and supported by multistate public opinion. Turner writes,

> I shall recognize as tests of sectionalism all of those methods by which a given area resists national uniformity, whether by mere opposition in public opinion or the part of a considerable area, or by formal protest, or by combining its votes in Congress and in presidential elections and also those manifestations of economic and social separateness involved in the existence in a given region of a set of fundamental assumptions, a mental and emotional attitude which segregates the section from other sections or from the na-

> tion as a whole. Sooner or later such sectional influences find expression in politics and legislation. . . .[13]

The governors have lodged multistate protests on a number of matters related to energy development, there is considerable opposition in public opinion in the West to unreasonable and unplanned industrialization, and there is a distinctive Western culture, attitude, and life-style that pervades most of these states. What Turner regards as sectionalism may become identifiable in the future if Western governors' actions are but the first of an evolving series of multistate actions.

Analysis of four distinctions drawn by Odum and Moore between sectionalism and regionalism in a Western context helps us arrive at some conclusions. First, regionalism is "a cooperative concern," whereas sectionalism has an "essentially competitive emphasis." With respect to their collective actions, the governors in their attitude are more cooperative than competitive, although competition between the states remains. Second, sectionalism "emphasizes the autonomy inherent in political [state] boundaries," whereas regionalism is marked by "a cooperating group-of-states." The attitudes of the governors emphasize both these positions simultaneously. Third, sectionalism features "separateness," whereas regionalism "connotes component and constituent parts of the larger national culture." The intentions of the governors are more clearly in accordance with the latter. Fourth, according to Odum and Moore, there is the "implication of more of the designed and planned society" in regionalism than in the individualistic sectionalism.[14] The Western governors have stressed the need for thoughtful planning in energy matters both at the national and regional level.

From this checklist of indicators, it would seem that the posture of Western governors, as represented by the objectives and resolutions in WGREPO and WESTPO, is principally one of regionalism but with a strong underlying emphasis on joint state action rather than regional action philosophically conceived.

Appendix:
Regionalism and Regional Institutions as Subjects for Political Analysis

By 1974 a number of Western states had experienced the early phases of massive energy development programs drafted and promoted by the federal government, utilities, and private energy companies. The governors of these states registered a consensus of firm and widespread dissatisfaction with many of the problems caused by rapid growth and energy development. The conflict has resulted in the cooperation and solidarity of this bloc of states as evidenced by the establishment of regional energy policy organizations. These events present interesting questions to students of politics and government. Is the multistate movement an expression of sectionalism or regionalism?

The meanings of *regionalism* and *sectionalism* are difficult to differentiate. Scholars have used the terms but have seldom precisely defined them. During the course of this appendix, various uses of the terms will be examined; the aim is to achieve some relatively precise differentiation. The position taken in this study is that the multistate response to energy development in the

West is symptomatic of certain aspects of both regionalism and sectionalism. The states have not become sectional to the extent that the North and the South were during the Civil War, but sectional feeling surrounds the view that the West's interests are in conflict with those of the East. The defensive posture that has developed against the external pressures for indiscriminate development of energy has been expressed in regionwide actions, but the tone of separatism associated with sectionalism appears to be lacking.

This book has examined within a regional context the motives of the creators of the regional organizations, the functions they expect the organizations to fulfill, the degree of commitment of each of the different states to the movement, and the degree to which energy problems will direct the functions of the organizations. The specific objective of this appendix is to examine the American experience of regionalism and to describe the approach that was taken in examining the multistate movement and the formation of regional energy organizations.

THE REGION: CONCEPT, MEANING, UNIT OF ANALYSIS

In order to make a valid judgment on the sectional or regional nature of the multistate movement that WGREPO and WESTPO symbolize, the evolution of uses of the concepts requires understanding. In addition, it is appropriate to analyze the views of a selection of prominent scholars of regional phenomena. Both these undertakings enhance the capacity to evaluate the state of the political West today as section or region. As a unit of analysis, the region has peculiarities and limitations that have received considerable examination by scholars in a number of disciplines. An awareness of the advantages and disadvantages of regional studies, particularly in regard to political phenomena, is advisable.

As Felix Frankfurter pointed out, the concept of region has meant different things to different disciplines, government administrators, and leaders in industry, and the nature of a region varies with the needs, purposes, and standards of those using the concept.[1] Just as the nature of the region is not stable, and as interpretations of its boundaries vary, so the meaning of the concept has evolved throughout the history of the United States. It is appropriate at this point to review the evolution of the concept and to consider the points at which its meaning has changed.

By the twentieth century, the United States had been divided into areas of "homogeneous diversities" in scores of different ways by government officials and by scholars. The terms used to identify areal sets included *parts*, *quarters*, *sections*, *divisions*, *regions*, *subdivisions*, *subregions*, *areas*, and *districts*. But the point to note is that *section* was the most commonly used term between about 1750 and 1900, and that *region* did not become a widely recognized term until this century. Before 1895 a *region* was generally agreed to be within a *section*, but in any event was seldom used as an areal concept. By 1750 British America was thought to be sectionalized, and the main criteria used to characterize a section were differences in situation, climate, economy, society, inland areas, tides, terrain, crops, and people.[2] The development of the sectional concept is mainly attributed to geographers, and when the Continental Congress convened, delegates used the term to refer to the three principal sections of the era—Eastern, Middle, and Southern. Sectional administration was adopted for naval, military, and Indian affairs, and by 1803 legislators in Congress had begun to add Western states as a fourth great division of the country.[3]

As we have seen, the words *section* and *sectional* received a new interpretation in 1806, when Josiah Quincy, speaking in the House of Representatives, warned that "whenever any great section of the Union shall deem itself neglected and the opinion becomes general among the people that they are sacrificed or abandoned; that they have not any, or their just weight, in the national scale, a series of struggles must commence, which will terminate either in redress or in convulsions."[4] The use of *section* in this new sense gained popularity in legislative and public circles as struggles for power and interstate conflicts produced crisis. The Civil War, the nation's greatest "sectional" conflict, probably stimulated the use of the concept, as even geographers turned to using the term increasingly up to 1900 after favoring *division* for so long. It would appear that Sir William Craigie's definition of "sectional feeling" as "a feeling aroused between one's own interests and the interests of people in other sections"[5] was appropriate to the conflicts of the mid nineteenth century.

The realities of the section in the United States were brought into public focus not only by politicians and political events but also by geographers, census officials, and historians throughout the period 1750–1900. Jedediah Morse, often known as the father of American geography, spearheaded an intellectual movement that refined the concept over the course of a century up to 1890. Morse used criteria such as soils, physiography, situation and climate, agriculture, commerce, manufacture, racial stocks, religious membership, and political beliefs to characterize sections. The compilers of the decennial U.S. census arranged their statistics and indices in sectional frameworks that evolved as research and data collection techniques advanced. Historians, notably a group associated with the University of Wisconsin, had been advancing the scientific study of sections in American history in the late nineteenth century. Wisconsin historian Humphrey Desmond claimed in 1888 that "the controlling fact in American politics was the fact of sectionalism."[6] The most influential scholar of history to focus on sections, Frederick Jackson Turner, brought widespread recognition to this approach in American historical studies. Using indices made up from physiographic, economic, institutional, and population factors to show the character of sections, and with references to the life and spirit of these sections, Turner guided the application of the concept of sections to fruitful historical research.

At the beginning of the twentieth century, the study of sections that made up American geography and history and the use of the terms *section* and *sectional* were pervasive. But with the publication of a monograph in 1895, Maj. John Wesley Powell, head of the United States Geological Survey, introduced a classification of American geographic "regions," a term that was to increase rapidly in currency up to the present time. Even though the census of 1910 was based on three great sections—North, South, and West—subdivided into nine divisions, and even though Frederick Jackson Turner argued for a six-section nation as late as 1935, the use of regional nomenclature spread rapidly. The first third of the century was a period of growing recognition of the region as a useful economic, administrative, social, and natural area in which various public services could be focused.[7]

During the 1930s regionalism increased in popularity as a concept for scholars and government agencies. The creation of the Natural Resources Planning Board in 1934, the creation of the Tennessee Valley Authority as a multistate regional authority, and the writings of Lewis Mumford[8] and Howard Odum and Harry Moore[9] during the 1930s helped spread the use of the concept. The years after World War II and up to 1979 saw the regional concept applied increasingly to natural regions, such as river basins, for planning, and to economically depressed regions for development and economic stimulation. Institutions established by these planning and development programs have commonly been joint state-federal arrangements. The regional movement has evolved to include Federal Regional Councils, regional divisions of private companies, proposals for land use and environmental management in regionwide government programs, and even a proposal by Rexford Tugwell for the adoption of regional republics as replacements for states.[10]

We have briefly described some aspects of the general evolutionary process of the concepts of section and region. A closer examination of the views of prominent regional scholars permits a more detailed understanding of the meanings of the terms and their similarities and differences, if any. Rupert Vance put forward a dynamic theory of historical evolution in the United States that provides a practical analytic framework within which to consider the meanings of regionalism. In effect, Vance argued that American history, in political and economic terms at least, is continually maturing from an era of frontiers to an era of social control. "The transition from an unsettled country marks the gradual change from the frontier process to social control. The intermediate stage in the transition is sectionalism; the ultimate stage may well be regionalism and regional-national planning. . . . our great major areas pass from frontier, to sections, to regions."[11] The thrust of this view, in part, is that sectionalism was basic to the evolution of regionalism rather than coincidental with it.

Frederick Jackson Turner, whose contribution to the understanding of sectionalism in American history has been most significant, developed criteria for measuring sectionalism. "I shall recognize as tests of sectionalism . . . methods by which a given area resists national uniformity . . . opposition in public opinion . . . formal protest . . . combining its votes in Congress and in presidential elections, and also those manifestations of economic and social separateness involved in the existence in a given region or a set of fundamental assumptions, a mental and emotional attitude which segregates the section from other sections or from the nation as a whole. Sooner or later such sectional influences find expression in politics and legislation, and they are even potential bases for forcible resistance."[12] Whereas Vance and Odum and Moore have contended that sections and sectionalism were steps in the process of cultural evolution and will be superseded by regionalism, Donald Davidson has argued that when regionalism is tested by political arrangements it is still sectionalism. Historical and structural sectionalism is so ingrained that it may always be present.[13] Davidson also feels the terms *sectionalism* and *regionalism* are far from being opposed to one another.

Some definitive research on the history and meaning of sectionalism and regionalism has been carried out by Odum and Moore. These scholars argue that sectionalism in the nation is analagous to nationalism in the world at large, each signifying isolation and self-sufficiency. Important distinctions between sectionalism and regionalism exist, according to Odum and Moore. First, regionalism places the nation first, whereas sectionalism sees the nation after the region. Second, where sectionalism features separateness, regionalism connotes component and constituent parts of the larger national culture. Third, sectionalism emphasizes the autonomy inherent in political boundaries and state sovereignties. Fourth, the regional concept implies a national cooperative process, which sectionalism does not.[14] Their meaning of regionalism—"a new American social economy and societal determinism"[15]—is, they argue forcefully, the opposite of the term's most common interpretation as a description of localism, sectionalism, and provincialism.

The foregoing analysis of the history and meanings of regions and sections in the American experience is necessary for an important reason. Regionalism, many scholars agree, connotes a different type of political posture from sectionalism. The different meanings attributed to the two terms must be clarified. Historically, it has been argued, the activities of sections in the

United States have become increasingly less distinct because of changes in economics, homogenization brought about through migratory patterns, the advances of communication technology, and other aspects of modern industrial society. Those who make this argument, especially Odum and Moore, suggest that regionalism reflects a cooperative societal determinism without symptoms of separatism. The movement in the West today, as exhibited in some statements of certain governors, may be symptomatic of both sectionalism and regionalism.

The regional concept has been very attractive to scholars in numerous disciplines and to government agencies. As a tool for research or as a unit of analysis, regionalism continues to be used frequently, but is associated with considerable disagreement in definition, concept, and structure. A general definition of *region* marks it as a unit of areal and cultural differentiation. Regional structure is capable of analysis both as a spatial pattern and as a fact of social organization. An almost unlimited array of concepts is used in the regional theories of the various disciplines. Although there is, similarly, a great array of conceptual frameworks in use in regional studies, the theoretical contributions of the field are as yet very limited. A significant factor, which in part determines the low level of theoretical content in regional studies, is that the choice and delineation of regions are functions of the research undertaken and of the objectives of the analyst.

As a unit of analysis, the region has been scrutinized by almost every branch of the social and physical sciences. Aspects of research of other disciplines that are of special interest to the political scientist would include the areal and environmental impacts on behavior *(Geopolitik)*, the effect of migrations and social inheritance on institutional arrangements in frontier areas (geographic determinism), the relation between the economy of a region and its government (political economy), and the political implications of interrelationships in spatial groupings of physical and societal phenomena described by regional scientists such as Walter Isard.[16] But generally it is fair to claim that political phenomena within regions have seldom been examined. As a tool for research and as a unit of analysis, the region has not attracted many students of politics.

APPROACHES TO REGIONAL PHENOMENA IN POLITICAL SCIENCE

A significant body of literature exists on two fields of politics related to regionalism. The first has focused on the politics of regions; the second has been concerned with the characteristics and performance of regional organizations.

Major contributions to the study of regional politics after 1950 are viewed here through an analytic framework resulting from the scholarship of Vincent Ostrom. Research on political phenomena in regions prior to 1950 is generally spread across the spectrum of interests of students of politics and is not easily classified. Nevertheless it may be profitable to report on the major areas of study and the most prominent scholars of regional phenomena in the early decades of the century.

Research on the political and administrative aspects of regionalism and regional organizations in the United States through the mid 1930s has been classified and reviewed by Odum and Moore[17] and the National Resources Committee.[18] The authors of the National Resources Committee study claimed, "The political scientist has quite commonly concerned himself with regions, but his interest seems to have lain not in the problem itself but in the question of what administrative use has been or can be made of some elementary form of the regional concept.[19] Writing three years later, Odum and Moore gave a more comprehensive account of the interests of students of the politics of regionalism. They showed the existence of scholarship in problems of areal government, the political aspects of sectionalism, the classical problems of federalism and states rights, and the administrative arrangements for regional matters.

Research on various aspects of regional factors in political theory had been going on in Europe and the United States for decades. Federal theories of the state, political development, regional conflicts, and political factors in sections and states were some of the most prominent areas of regional theory developed before the 1940s. The debate between supporters of local government and the promoters of centralization centered on areal and regional theory. Odum and Moore drew attention to the work of Linton on forms of government for areal units, the work of Dunning on physical determinism and political behavior, and the approach of Bowman and Huashofer to *Geopolitik*, or political geography.[20]

Before developing their own positions on the political and administrative dimensions of regions, Odum and Moore reviewed in detail some of the existing research efforts, at that time, in the politics of regionalism. William Munro, in publications during 1928, called attention to social problems too big for any single state, yet not big enough for the nation, that should be addressed by regional government.[21] Marshall Dimock argued that in certain problem areas there is need for administrative coordination between states and the federal government.[22] In a book written in 1935, William Elliott generally supported Munro's argument calling for new sovereign regional institutions in the United States.[23] Charles Merriam and John Gaus wrote strongly in favor of regional institutions as an appropriate and necessary form of organization for government in the United States,[24] and James Fesler and Donald Davidson wrote on various aspects of administrative regionalism as well.[25] The writings of Lewis Mumford in *The Culture of Cities*,[26] and of a number of students of politics and public administration led by Gaus, Dimock, and Merriam in their association with the National Resources Committee each added to the attractiveness of regional solutions to problems of natural resources and public affairs stretched between state and nation.

Research by political scientists on concepts and theory in regional problems continued during the 1940s, mostly concerned with administrative arrangements. The establishment of the Tennessee Valley Authority in 1933, of multistate regional planning agencies in New England and the Pacific Northwest, of the growth of regional administration programs in various functional areas, and of the establishment of urban and river basin regional organizations held most of the attention of students of politics and administration at that time. David Truman[27] and Kurt Wilk[28] wrote on administrative regionalism in specific functional fields, such as agriculture, fiscal policy, Indian affairs, social security administration, natural resources conservation, and in particular river basin development.[29]

Two studies that were significant contributions to theory in areal and governmental power, one by Arthur Maass[30] and the other by James Fesler, appeared in 1939 and 1949. These works dealt with the concept of region in the context of natural areal groupings of men and the environments that conditioned their existence. Under the assumption that man's needs change and his relationships to the environment also change, Fesler drew attention to problems of redesigning governmental areas to fit particular functions. The alternative is to accept the existing areal framework and to readjust functional relationships among existing government areas. Fesler argued that "constitutional obstacles and sheer inertia all make regional governments an unlikely answer to the need for reciprocally adjusting areas and functions."[31]

Two other studies published during the 1940s focused on the politics of regions rather than administrative matters. Thomas Donnelly's work on Rocky Mountain politics[32] and V. O. Key's on Southern politics[33] were akin to the studies of politics in multistate areas that increased in numbers after 1950.

In 1956 a framework for the analysis of a regional political system was proposed by Vincent Ostrom. Ostrom developed five approaches that could be used by students of politics. His framework will be used a a guide here for our examination of the major studies of regional politics.

Ostrom pointed out that the study of a political system for a regional community, without institutional arrangements coordinate with its world, poses difficulties for analysts of regional political phenomena. Traditional institutional referents used by political scientists, such as government, legislatures, and organizations, are not adequate as a basis for investigation in a noncorporate region. With the absence of regional institutional arrangements as referents, Ostrom argued that a behavioral approach, through the analysis of policy formation, is more useful to the analyst. Using the policy approach, Ostrom argued that the "order of relations necessary to describe any political systems include: (1) definition of the appropriate world or universe of a region; (2) the population and its characteristics; (3) the organizational structure; (4) processes of action; and (5) the law and policies applicable to the region."[34]

The definition of the region to be examined, as previously noted, depends on the intentions of the analyst and the problems to be addressed. As Ostrom indicated, the scope of a region is determinable as the area encompassing the variables pertinent to the problem of concern to the analyst. This definition facilitates the delineation of whatever regional boundaries an analyst desires. An important point that Ostrom included in the definition of the region was the necessity to examine also political relations external to the bounds of the region itself. Ostrom's intention appears to have been to demonstrate that a region has a relationship to the whole nation and that regional political phenomena should be examined in the context of relationships with the larger entity or nation.

The second approach Ostrom put forward required the analyst to delineate the population of the region, the demographic characteristics, and the patterns of political behavior and to inquire into the existence of a regional consciousness or perspective. Policy determinations and actions within the region would be significantly influenced by the way the populace, and in particular the leadership elites, conceptualize problems and their solutions. Regionalism, philosophically conceived, is different from joint state action founded on parochialism, and either of these states of mind influences policy formation. An examination of the views and intentions of a representative selection of elites within a region would help the analyst to report on the extent of a regional consciousness. In addition, Ostrom pointed to the need to examine elites in a decision-making capacity external to the region, as well as that within the region.

To understand the "nature of a region and its potentialities," the analyst must examine the relevant political structure, Ostrom argued. This structure includes the diverse set of competing actors making demands upon subordinate (state and local) and superordinate (federal) agencies operating in the region. The principal actors in the movement to address energy policy problems on a regional scale in the West today are governors. The potentialities available to the governors in state agencies are severely limited, as energy problems cross political borders. The potentiality of federal agencies is limited too, as they are viewed as part of the regional problem. Therefore existing political structures, both state and federal, have limitations as decision-making instruments. According to Ostrom, the analyst should thus examine the activities in a region directed at structural reform or innovation.

In the fourth approach, Ostrom argued for the need to

examine the "mode of operation" in the regional system and the "patterns of action" between actors seeking to realize regional objectives. These patterns of action are believed by Ostrom to have consequences that affect potentialities for regional development. Long-standing state-federal patterns of action and state-state relationships would be major aspects of these patterns. Previous experiences of interlevel governmental relations, previous center-periphery relations (national-regional), and even the experience of a region with regional institutional arrangements would appear valid indicators of some aspects of the politics of regionalism.

Finally, a policy analysis approach was developed by Ostrom on the assumption that policies may be analyzed "in terms of their commitments to certain values and actions which can be expected to ensue." By examining the policies set by interests promoting regional action, the analyst can raise questions of the extent to which these interests conceive of the policies in terms of commitment to patterns of regional development. If the analyst discovered that regional policies were expressing a firm commitment to regionwide objectives, with little evidence of state parochialism or selfishness, he could expect a higher probability for the survival of a multistate policy organization.

THE POLITICS OF REGIONS

The studies of political and administrative phenomena in a regional context have been restricted here to those dealing with multistate regions. The concern with multistate regional literature is due only to our focus on ten Western states. The concepts and analytic techniques of this literature are appropriate to our concerns, but the literature of the politics of metropolitan regionalism or multinational regionalism may also be relevant.

The study of Southern politics by Key[35] set a pattern that was to be followed in other regions by political scientists. Key argued that the Southern states were culturally and politically distinct from the rest of the nation. (The states to which he referred were North Carolina, South Carolina, Tennessee, Georgia, Alabama, Florida, Virginia, Mississippi, Texas, Arkansas, and

Louisiana.) Several of Key's students, namely Duane Lockard,[36] John Fenton,[37] and Frank Munger,[38] approached the study of politics in other regions along similar lines.

The politics of the West has been studied in a regional context by Frank Jonas[39] and by Thomas Donnelly.[40] Daniel Elazar[41] has divided the nation into political culture regions. Ira Sharkansky, himself a contributor to the study of regional politics, points out that the regional literature dominated by this group of scholars is limited in scope and has focused almost exclusively on electoral processes, party competition, and intraparty alignments.[42] These studies will be examined for their objectives, the approach used, and their findings.

Fenton's study of the border states—Missouri, Kentucky, West Virginia, and Maryland—avoided using the terms *region* and *section,* but it did encompass an examination of the characteristics of political life within these contiguous states. Of primary concern were the political consequences of border states culture. Fenton argued that geographically, culturally, and climatically these were middle states, between the South and North, in which two population and cultural fronts met. The movement of population into the border states was believed to have created a similar political configuration in all four. Patterns of political organization and political change were found to be common to these states and were defined by Fenton in four aspects of political structure: the evolution of a three-party political pattern, a distinctive type of organization within the Democratic Party, similarities in candidates offered by leading parties, and a special type of political change. In addition, Fenton claimed that the South was destined to inherit the political and cultural model of the border states. Fenton's choice of only four states has been criticized, as, according to Sharkansky, Delaware and Oklahoma should also be included. Fenton's analysis of regional politics has not included concern for the debate about regionalism or sectionalism, or any attempts within the group of states to pursue multistate policies.

Lockard was not interested in the debate over the meaning, or even the relevance, of the term *region.* His purpose was to carry out a comparative analysis of state political systems within a regional framework in New England. Lockard evidently used a regional framework because it reduced uncertainties and variables to more manageable proportions.[43] In his examination of political parties and legislative politics over recent decades, Lockard discovered a number of similarities in the governmental structure of the states. He admitted the existence of two distinctively different political systems within the region but found the similarities to be plentiful and significant. Lockard pointed out such common structural features as town meetings, the nation's largest legislatures, strong political parties, the minor role of county government, the peculiarities of constitutions, and the similarities of political heritage within the group. Sharkansky, however, strongly criticized Lockard's choice of boundaries for the region, arguing for a separation of southern and northern New England, in part, because of differences in party competition and socioeconomic characteristics.

The purpose of Donnelly's edited work on Rocky Mountain politics was not to examine the definition of region or its applicability to states "but to bring their politics to the attention of students of political parties."[44] Donnelly used the terms *section* and *region* interchangeably, without drawing any distinction between the two. The eight states examined in his study were Montana, Idaho, Wyoming, Nevada, Utah, Colorado, Arizona, and New Mexico. Donnelly's was not so much a study of regional dimensions of politics in the West as a description of state politics in the region.[45] It added five more states—California, Oregon, Washington, Hawaii, and Alaska—to the definition of the West. The purpose of the study was not to argue for regional recognition of any combination of these states but to show, in the opinion of Jonas, that the thirteen states "can be viewed as a region with an underlying unity in physiographic features, economic impulses and political objectives."[46] This study, like a number of the others previously discussed, had the examination of political parties as a primary objective. But more than any other analysis at that time, it made an attempt to identify a political region. Potential causes of political unity and disunity in the region were identified as well as historically significant causes of unity and divisiveness.

A study of American state politics edited by Frank Munger has also contributed to the literature of the politics of regions. For purposes of examination, the South, Northeast, Midwest, and West were separated as specific regions. The primary purpose of the work was to permit comparisons between state political systems, but Munger's premise was that regional and sectional differences exist, producing state and regional political variations.[47] The political phenomena of each region were considered under the topics of electorates, parties, legislatures, governors, courts, and local governments, but the categories were generally divided among the states making up the region. In the West, for instance, where eleven states were grouped, Oregon was the case study for the legislature, Utah for the governorship, and the Colorado River for the courts. The studies, furthermore, were not synthesized within the region as an entity.

Other works that should be reviewed as part of the literature are Elazar's on political cultures and Sidney Tarrow's and Joseph Hilhorst's on center-periphery administrative relations and politics. Elazar argued that three main factors were responsible for shaping states' political structures, electoral behavior, and modes of organization for political action. These factors were defined as political culture, sectionalism, and the continuing frontier. Elazar made clear his interpretation of the differences between regionalism and sectionalism, arguing that the national political culture is a synthesis of three major political subcultures and that each of these subcultures is found in specific sections of the nation. The political subcultures, known as "moralistic," "individualistic," and "traditionalistic," have a geographic distribution, and according to Elazar, their distribution has founded sectionalism.[48]

Tarrow's work focused on France and Italy and provided a useful approach to the study of political and administrative phenomena in regions. In his study of regional reform in certain nation-states, Tarrow examined the issues of area and power, inducers and inhibitors to regional reform, and the influences of institutional setting and various actors on the process of reform. While not a study of the dynamics of all the phenomena of a regional political system, Tarrow's approach is practical, as it brings into focus the actors and institutions central to the development of regionalism.[49]

A theoretical approach to the study of center-periphery relations, or the relationship between region and nation, was developed by Hilhorst. Based on a concept of domination, some areas of a nation may be geared to or led by decisions made in another area. Often the center (national government) may dominate the periphery (a state or group of states). The power that the center holds over the periphery may be due to its greater population, size, or bargaining power, or its access to superior information. In political terms, the superiority of the center over the periphery is expressed by constitutional authority, preemptory rights, areas of public policy, and voting dominance in Congress. The center's domination over the periphery may exhibit an extractive component in situations where the center gains a profit from relations with the periphery. The periphery may reduce the domination so that the center has only regulatory command. The more the periphery can gain in relative bargaining power, the less the center can use center-periphery relations to its advantage. As Hilhorst suggests, the more access the regional or peripheral elites have to information, the more input they have to relevant decision making. As the sophistication of regional elites increases, the values of periphery and center may gradually overlap.[50]

The most analytically precise and definitive work on the politics of American regionalism, Sharkansky's *Regionalism in American Politics* (1970), will be the last sample of the literature to be examined here before we turn to the literature on regional organization. Sharkansky set forth the viability of regionalism in American politics. "It leaves an imprint on the character of political campaigns and the nature of services that citizens receive from their governments."[51] He criticized the regional literature for drawing questionable conclusions about the distinctiveness of politics of a region, for failing to consider alternative definitions of a region, for failing to sort political from other determinants, and for failing to explore a wide range of factors systematically on a regional basis and on intraregional and interregional bases. In an effort to circumvent these shortcomings in his own study, Sharkansky examined the forty-eight contiguous states, defined seventeen regional groupings of states, attempted to separate the political explanations of regional phenomena from the sociocultural and economic, and turned to the quantifiable aspects of political phenomena. In insisting on the use of quantifiable measures of politics and policy (sixty-one measures in total), Sharkansky was aware of the cost of excluding certain aspects of politics.

The most prominent political science literature prior to Sharkansky's work was limited at that time to six regions—New England, the South, the Midwest, the West, the Rocky Mountain states, and border states. But Sharkansky cogently demonstrated that regional borders shift with the aspects of politics that are at issue. His definition of the region is precise and leads into one of the major findings of his study:

> Region is a diffuse aggregate of economic, social, and political phenomena that include cultural predispositions to particular forms of public service or political relationships; migrations that have transferred practices from one section of the country to another; war and military occupation that have helped to mold nascent prejudices into firm principles of behavior; and the tendency of public officials to select their own norms from those of nearby states. The character and relative importance of these ingredients of "regionalism" vary from one region to another.[52]

In his first effort to separate the economic determinants of regional characteristics from the cultural and the historical, the behavior of state elites became a major focus. He found a tendency for state elites to consult with their counterparts in neighboring states to set regional norms. He concluded, "What does appear from region to region in this examination is not the power of a single explanation for regional peculiarities, but the power of regional norms to persist over time and to influence current styles in state politics and public services."[53]

A review of regional political phenomena is appropriate here. Of importance is to what degree there are obvious gaps in approaches that could fruitfully be used by students of politics.

The first consideration to the analyst of politics within regions, according to Ostrom, should be the definition of the region itself. The studies examined above collectively placed little emphasis on justifying their definitions of regions, the exceptions being Jonas and Sharkansky. Little or no attention was given to the issues of boundaries for regions. Both Sharkansky and Elazar investigated the boundary issue in detail. Ostrom and Sharkansky emphasized the need to evaluate political phenomena in the context of the region being part of a greater whole, but this aspect has not been of principal concern to students of politics. Even though Sharkansky as well as Ostrom pointed out that the definition of a region is up to the analyst, and therefore flexible, there was a noticeable lack of concern for describing the bases for boundaries.

Ostrom's second approach emphasized the description of a region's demographic characteristics and patterns of political behavior. Such descriptions were found in varying detail in the studies of Fenton, Lockard, Munger, Jonas, Sharkansky, and particularly Elazar. But no study undertook to measure the existence of a regional consciousness in the people of the region. In addition, both Ostrom and Sharkansky underscored the importance of the perspective of regionalism held by elites. No studies to date have undertaken this exercise either. Ostrom took this matter a step further and called for analysis of the elites external to the region who may have influence upon it. This too has received no full treatment. In addition, there is little evidence of analysis of "regional norms" that evolve between elites in contiguous states.[54]

Analysis of the organizational structure of a region was the third approach suggested by Ostrom. Munger examined a number of aspects of the structure within states but not of the region as a whole. Fenton and Lockard examined parts of the organizational structure of separate states, but the task of analysis of regional movements or regional structures remains unattended. In effect Ostrom's suggestions have not been heeded. Hilhorst's theoretical piece on center-periphery dominance did, however, provide a perspective that analysts might use to explain some of the political phenomena surrounding regional movements.

Next in Ostrom's five-part approach was the "patterns of action" that emanate from actions trying to realize regional objectives. Some of these actions have been interpreted here as previous governmental experience in a region to establish multistate mechanisms or to support regional activities. Jonas's study in 1969 did examine, in brief, the historical experience of the West with regionalism, notably in the contribution by Neal Maxwell. But the literature generally is devoid of systematic and scholarly analysis of the patterns of action between principal actors in a region, or between levels or agencies of government. Nor is the evolution of regional actions and mechanisms anywhere discussed. Aspects of the structural and institutional phenomena of regions have been addressed in the literature; an examination of these follows in the next section.

Finally, Ostrom's fifth approach favored the examination of policies drafted for a region. Ostrom suggested in addition that there was need for analysis of the intentions of the creators, both individuals and organizations, of such policies. Policy analysis in the area of regional decision making and law has received scant attention in the literature reviewed above. This approach will be seen as part—although usually incidental to other foci—of some studies that have addressed regional organizations.

STUDIES OF REGIONAL ORGANIZATIONS

The study of regional organizations has gone largely untouched by political scientists. The majority of the few studies that have been done focus on interstate compacts, and very little indeed is known of the other types, described below in Hartley's typology.[55] In part, the lack of interest in these organizations may be explained by the fact that the federal system makes constitutional provisions for only two sovereign levels of government. Neither in an operational nor a legal sense do regional units fit into this system. But it is true that the regional concept persists. Several political scientists have agceed with Richard Leach that "regionalism must be recognized as a major new development in modern American federalism."[56] As the number of regional institutional responses to problems of public policy appears to be growing rapidly in this second half of the twentieth century, institutions have become more attractive subjects for political analysis. We now turn to an examination of the major studies of regional organizations, with particular interest in who it is that creates regional organizations, and what their purposes are.

Regional organizations have been created by any one of five means: interstate compact, parallel state legislation, federal legislation, federal and state legislation in combination, and executive action at the state or federal level. The Constitution of the United States makes provision for two methods of adjustment for problems between states. One is the grant to the Supreme

Court to rule on controversies between states, and the other is expressed in the Compact Clause in the Constitution. There are two means by which interstate compacts may be negotiated. One is by reciprocal legislation, and the other is by the contract system.[57] Interstate compacts are the most common form of regional organization in the United States. Weldon Barton subdivided compacts into four categories, mainly on the basis of function: regulatory compacts with advisory or operational capacity, metropolitan interstate compacts, river basin development compacts, and state service compacts.

As there are different ways of creating regional organizations, so are there different forms in existence. David Hartley has identified six main types of regional organization, and subtypes within some categories. The first type is the economic development institution such as Title V and the Tennessee Valley Authority, which are both federal-state partnerships. The second type is the river basin regional institution in which states are joined by interstate compact; it is also a separate subtype of joint federal-state organization designed for river basin planning known as Title II. Standard federal regions, the third category recognized by Hartley, are not formal organizations but are arrangements which have led to the formation of federal regional councils. These councils are principally discussion and coordination instruments for the purpose of simplifying and coordinating field operations of the major federal departments. State-initiated regional organizations comprise a number of subtypes: the compact; national and regional organizations serving components of state government such as the Council of State Governments; the governmental–private enterprise nonprofit corporation represented by the unique Federation of Rocky Mountain States; the executive-initiated model created by fifteen governors in the South known as the Southern Growth Policies Board and supported by member states; and the executive-initiated model created by ten governors in the West known as the Western Governors' Regional Energy Policy Office supported by federal Title V funds. The last two types encompass a variety of nongovernmental and interstate metropolitan and rural regional organizations.

In an examination of multistate region organizations, Martha Derthick developed the categories of "forums of peers" and "catalysts." Operating managerial or regulatory bodies such as the Tennessee Valley Authority or the Delaware River Basin Commission were classified as forums where various organizations are represented whose activities overlap. The organizations involved in planning and coordination, such as the Appalachian Regional Commission and Title V and II commissions, were classified as catalysts because part of their activity is defining goals for other organizations to pursue in the region.[58]

Marian Ridgeway, in her examination of interstate compacts, analyzed four compacts involving the state of Illinois. She probed their history and sought a definition of their purposes. Her research showed that governors are among the most enthusiastic supporters of interstate compacts, and she called for far more intensive research on this phenomenon. Ridgeway speculated that the reasons for gubernatorial favor of these institutions could be one or all of the following four: through them a governor may gain more power over state government or avail his state of federal funding or the resources of other states; some social problems such as poverty or energy policy are easier to manage on a multistate basis; regional institutions can be useful to defend states' rights against the threat of federal preemption in some aspect of policy; and as he appoints his state's representative to the regional body, the governor may gain some leverage over state bureaucracy if the appointee reports directly to him. Ridgeway was keenly sensitive to the limitations of the compact instrument and was critical of certain design features and practices. "What these compacts reveal," she remarked, "is a state acknowledgment of need in a federal system, but a political response that appears to be little more than a state gesture along the lines of self-protection rather than towards responsible and constructive achievement to eradicate the problems that created the need."[59]

A similar criticism of interstate compacts was rendered by Charles McKinley in 1955 when he argued that "the interstate compact contrivance thus far suggested for meeting the need for regional public policy formation and administration of land and water resources appears to be a cumbersome, jerry-built structure lacking in regionwide political responsibility, parasites on national finance, and negative or unduly dilatory in decision making."[60]

It was McKinley's impression that members of regional organizations did not feel regionwide loyalties. He held that the recrudescence of states' rights, one of the causes of regional organization, was a sad anachronism. One might counter, however, that if in an area of public policy, such as national energy policy, the federal government does not fulfill its responsibilities to the public and causes undue hardship to a group of states, the establishment of a regional mechanism with a defensive or constructive posture is warranted. In his paper McKinley quoted from a study by Leuchtenburg that identified a source of initiative for regional organization that has been commonly cited in the literature.[61] This source was private economic interests, which see regional instruments as means to forestall federal regulatory involvement in policy areas the interest groups wish to exploit.

An examination of Vincent Thursby's analysis of interstate compacts reveals key political and administrative problems that could potentially prompt states to form regional organizations. The first is the problem of scope. As Thursby points out, public policy that does not coincide with state lines can be troublesome. Water problems in the West are a good example, as states that share the same river cannot afford, for the sake of administrative effectiveness, to act in isolation. A second problem is based on the assumption that the national economy is really a sectional economy in disguise, and that forces of political regionalism jockeying for power may create protective boundaries such as tariffs or organizations to promote regional interest. A state, or more than one state, may agitate for regional organization when it finds its interest, such as in air quality, threatened by inaction or lax administration in a neighboring state.

The experience of interstate compact agencies at work was the subject of analysis in a study by Richard Leach and Reading Sugg. Six case studies led to some interesting conclusions about the reasons for this form of regional organization and the intentions of their creators. The interstate compacts, according to Leach and Sugg, "must be regarded as the carefully selected tools of energetic states to exert their powers effectively." Generally, the authors note, the party states have not regarded the compact agencies as regional units distinct from themselves. "The states plainly regard them as instrumentalities of state power that enable the states to work together in reaching particular goals. The states have made it clear that they dislike the prospects of independently powerful regional authority as much as greater federal control." Leach and Sugg argue that states have been "using compacts to create agencies for joint 'state' action rather than for regional action philosophically conceived."[62]

In another study, Leach's analysis of the Interstate Oil Compact revealed that governors and private oil interests pressed for the establishment of a regional organization. Begun as an organization to serve a group of contiguous states, it has since evolved to include forty states across the country. The purpose of the organization was to influence states to conserve oil. Conservation was its practical mission, but the organization was founded on the philosophy of state sovereignty, Leach discovered. Formed to block federal encroachment on state sovereignty, the organization has worked as a watchdog for the interests of oil-producing states in Congress and before federal regulatory agencies. As a fact-finding and promotional body with the powers of persuasion only, it has been judged one of the nation's most successful and influential multistate organizations.[63]

One of the few studies of political aspects of regional organizations other than interstate compact agencies was carried out by Derthick.[64] She compared the results of different types of organizational response to the need for coordination and the effectiveness of these organizations in reducing conflict and inconsistency among federal executive agencies. It was Derthick's position that most regional organizations have been improvised. The forms and functions they assume are influenced by opportunities of the moment, by the idealism and interests of their founders, and by the political conditions at the time of their establishment. She argued cogently that the Tennessee Valley Authority was an accident and a by-product of issue and opportunity. The Delaware River Basin Commission was a means by which states could settle disputes without federal interference, at the same time simplifying relations with the federal government; as all the founders but one were Democrats, the fortunes of electoral politics helped its creation. The Appalachian Regional Commission began as a regional effort but quickly became involved in a pattern of defensive and selfish activities between separate states.

The Title V economic development agencies were found to be almost devoid of regional sentiment, and were attractive to state governors primarily as conduits for federal funds. Investigations have shown states to have little interest in Title II river basin planning agencies; they appear to participate in them only for reasons of defense against other states, the federal government, and other regions. It has been Derthick's conclusion that none of the regional organizations she examined grew out of a widely shared sense of regional interest. Only in special circumstances where leadership, opportunity, and political backing coincide may these organizations survive and carry out a significant share of public functions.

Examining a range of types of regional organizations as Derthick did, the Advisory Commission on Intergovernmental Relations made a major contribution to the knowledge of their legal, political, historical, and operational aspects.[65] This study focused on the legal base, organization, and operations of the Appalachian Regional Commission, five Title V agencies, five Title II agencies, and two federal interstate river basin compact agencies. Each regional organization examined was a federal multistate instrument, so other types of regional organizations were not considered. The main objective of the analysis was to determine if institutional experiments should continue, and if so what the best structural arrangements would be.

The commission concluded that not much was known of the organizations that had been in existence for such a short time, and that they should be continued since they looked promising. Unfortunately this was little more than Odum and Moore had concluded in the 1930s. The commission referred more particularly to federal multistate organizations that had proliferated during the 1960s, such as the Title V and Title II agencies, as the result of efforts to improve on traditional compact approaches which at times were viewed as inadequate and inappropriate. Conflict resolution between states, the promotion of states' rights, and administrative strategies to cope with problems of scale and coordination were each documented as inducers of regional organizations.

The commission's report pointed out the significance of the source of initiative for a regional program. This argument was developed cogently in a study of interstate compact organizations by

Weldon Barton, who recognized that fundamental aspects of regional organizations, and major determinants of their operation, are the intentions of the creators and the politics of creation. Barton also testified to the significance of the environment in which a regional organization is created.[66] This last point was more fully developed by Joan Aron[67] and by Robert Warren and Geoffrey Wandesforde-Smith[68] in their examinations of regional organizations. Roscoe Martin's writing on the inducers and inhibitors of regionalism and regional organizations dealt with aspects of both the politics of creation and the political-administrative environment in which organizations must function.[69]

In his analysis of several case studies of interstate compacts, Barton was principally concerned with the creators and purposes of these institutions. He investigated the politics of their creation under the assumption that these organizations were established with the expectation that benefits would accrue to interest groups or the public at large. He found that the Interstate Oil Compact Commission was established by oil interests and governors as a balk against federal encroachment on states' rights. It was clear to Barton that the Atlantic States Marine Fisheries Commission was a mechanism used by states and private interests to combat expanding judicial assertion of the doctrine of federal preemption. Private power and agricultural interests united to promote development of the Colorado River in a effort to persuade federal agencies to build Boulder Canyon Dam for power and storage purposes. Barton argued also that the Great Lakes Commission had served its party of states as a lobbying agency promoting development in the region. He also made the interesting observation that states may invoke interstate compacts for the purpose of disrupting direct national-local relations.

The politics of creation was also treated by Aron in her study of metropolitan regionalism in New York and the history of the Metropolitan Regional Council. A situation would have to approach a state of crisis in public policy before traditional government units would be compelled to joint action on regional problems, she found. Apparently the Metropolitan Regional Council evolved as an institutional approach to the problems of geographic scope of many government services within the metropolitan area. Aron's most valuable contribution lies in her identification of the inducers and inhibitors of regional action and the roles a regional organization must play to survive in its environment.

There is considerable overlap and agreement in the findings and propositions derivative from the works of Aron, Wandesforde-Smith, and Martin on the environment in which regional organizations have to survive. Some of the principal determinants of survival are related to the character of the region, the jurisdictions involved, the personnel in elite capacities, the degree of support received from other levels of government, the strength of regional consciousness in the public, the strength of competing regional forces, and the strategies pursued by the regional organization itself. As such, the main difficulties for the Metropolitan Regional Council were found to be public apathy toward intergovernmental cooperation, lack of consensus on a crisis that would induce regional cooperation, the antagonistic posture of the new institution toward other agencies, lack of interest in regionalism by state and local agencies and political parties, and the lack of authority and sense of purpose of the organization.

Broad propositions about regional organizations suggested by Martin focused to a considerable extent on issues and factors raised by Aron. The main inhibitor to regional action appears to be the absence of regional awareness on the part of the public. A region's social and political composition is more likely to promote dissonance than unity of purpose, and the hostility to significant change of those in positions of authority can work powerfully against regional action. Like Derthick, Martin felt that crisis and catastrophe, fortuitous political circumstances, and outstanding leadership were some of the chief factors in an atmosphere favorable to regional action, but were seldom present.

The inducers and inhibitors of regional action have direct bearing on the creation and operation of regional institutions. Warren and Wandesforde-Smith analyzed and evaluated regional arrangements for water policy making, focusing in part on the design of organizations, their organizational and operating patterns, and the political-administrative environment in which they function. This environment included public and private organizations, actors who would have impact on the organization's operation, and regional factors that might facilitate or inhibit the achievement of its objectives.

CONCLUSION AND PROPOSITIONS

From the analysis of the literature pertaining to regionalism and regional organizations, it is apparent that students of politics have made relatively few contributions. Public affairs in such areas as water management and energy policy transcend traditional political boundaries. The rapid growth in the number of regional organizations in these areas of public affairs should prompt more political scientists to undertake examinations of the trend. With some obvious confusion over the differences between sectionalism and regionalism, an inquiry into the exact nature of the multistate coalition in the West today is appropriate. By examining the multistate movement that has culminated in the formation of WGREPO and WESTPO, I have tried to contribute to our understanding of the politics of regionalism and regional mechanisms.

The analytical approach defined by Ostrom has been a useful measure for organizing and reporting on the literature of the politics of regions. The five-part approach, if adopted in full, would require a study of enormous proportions. Such a study was not intended here. But the approach has revealed some gaps in the literature that this study has addressed.

The analysis of the literature has revealed a considerable amount of agreement in different studies. The principal information sought in the analysis was an identification of the intitiators of regional organizations and their motivations. Major findings are as follows: first, knowledge of the creators' or initiators' intentions and understanding of the politics of creation are necessary exercises. Such a finding led to the proposition that *the motives of the creators of a regional organization and the politics of its creation are fundamental determinants of the organization's structure and objectives.*

Second, states' rights, state and regional defense in a supposedly sectionally oriented economy, and attempts by the periphery to reduce dominance from the center may each be common elements in regional movements. This understanding led to the proposition that *concern for external intrusion into energy development reinforces latent commitments to states' rights and long-standing hostilities to Eastern and federal influences.*

Third, inhibitors to regional action generally outweigh inducers. Several scholars have attested the lack of a widely shared sense of regional interest in the areas of their studies. In the last chapter we reported on the findings of a survey of a representative selection of elites in the region as to their understanding of the objectives of WGREPO and WESTPO and the aspirations of the governors. In addition, we attempted to ascertain from respondents the major inducers and inhibitors of regionalism in the West. The proposition testable by the findings of the survey is that *political factors in the West have tended to inhibit rather than induce regional action but the energy issue has at least temporarily stimulated the assertion of political regionalism.*

Notes

A note on sources

Because of the nature of the research for this book, a traditional bibliography is inappropriate. Sources, therefore, are cited only in the notes.

Chapter 1

1. Norman C. Miller, "Environmental Impact of Coal, Shale Mining Is Issue In The Far West," *Wall Street Journal*, 24 October 1974, p. 1.
2. David S. Broder, "Contemporary: Mountain West Allying Against Energy," *Chicago Sun-Times*, 24 November 1974.
3. Helena Huntington Smith, "The Winning of the West," *Washington Post*, January 1975.
4. "The Mountain States: Cooling the Boom from Energy Resources," *Business Week*, 27 January 1975, pp. 108–9.
5. Broder, "Contemporary," p. 12.
6. Eric McCrossen, "New Mexico in Key Position: Next 'Civil War' Issue: Energy," *Albuquerque Journal*, 17 April 1975.
7. "Coordination Need Cited in Western Energy Push," *Irvin Fall [Idaho] Times News*, February 1975.
8. "Western Governors Eye Ag-Energy Co-Existence," *Record Stockman* (Denver), 10 April 1975.
9. Mel Reisner, "Not Nation's Energy Colony," *North Wyoming News* (Worland, Wyo.), 19 April 1975.
10. "Western Governors."
11. *Balanced Growth and Economic Development: A Western White Paper*, (Denver: Western Governors' Policy Office, 1978).
12. See L. K. Caldwell, "Environmental Policy As A Catalyst of Institutional Change," *American Behavioral Scientist*, May–June 1974, pp. 711–30, "Energy and The Structure of Social Institutions," *Human Ecology* 4 (1976): 31–35, and "Future Policy Directions," in *Energy Impacts on Public Policy*, ed. Walter F. Scheffer (Norman: Science and Public Policy Programs University of Oklahoma, 1974), pp. 203–22; Don Kash et al., *Our Energy Future* (Norman: Science and Public Policy Program, University of Oklahoma, 1976); David W. Orr and Stuart Hill, "Leviathan, The Open Society, And The Crisis of Ecology," *Western Political Quarterly* 31 (1978): 457–69.

Chapter 2

1. Thadis W. Box, *Rehabilitation Potential of Western Coal Lands* (Cambridge, Mass.: Ballinger, 1974), p. 22. See also S. David Freeman, *Energy: The New Era* (New York: Vintage Books, 1974), p. 77.

2. "Energy: A Question of Survival," *Newsweek*, 26 August 1974, p. 65.

3. David S. Broder, "Mountain West Allying Against Energy Rape?" *Chicago Sun Times*, 24 November 1974.

4. Council on Economic Priorities, *Leased and Lost: A Study of Public and Indian Coal Leasing in the West*, vol. 5, no. 2 (New York, 1974).

5. Federation of Rocky Mountain States, *Energy Development in the Rocky Mountain Region: Goals and Concerns* (Denver, 1975), p. 21.

6. Irvin White et al., *The Coal and Oil Shale Resource Development System* (Norman: University of Oklahoma, Interim Report, 27 July 1974).

7. Box, *Rehabilitation Potential*, p. 26.

8. K. Ross Toole, *The Rape of The Great Plains: Northwest America, Cattle and Coal* (Boston: Atlantic, Little Brown, 1976).

9. John L. McCormick, *Facts about Coal in the United States* (Washington, D.C.: Environmental Policy Center, 1974), p. 1.

10. Box, "Rehabilitation Potential," p. 230.

11. Northern Great Plains Resource Program, *National and Regional Energy Considerations Work Group Report* (Demver, 1974), part 7, p. 42.

12. White et al., *Coal and Oil Shale Resource Development System*, p. 230.

13. Federation of Rocky Mountain States, *Energy Development*, p. 21.

14. Freeman, *Energy*, p. 98.

15. "Energy," p. 68.

16. Terry Noble Fiske, "Rush To The Rockies: Some Aspects of Mineral Development of Non-Fee Land," *Rocky Mountain Mineral Law Review* 7 (1969):39.

17. "Oil Field Drilling Halt Asked," *Denver Post*, 2 December 1977.

18. Federation of Rocky Mountain States, *Energy Development*, p. 21.

19. Northern Great Plains Resource Program, *National and Regional Energy*, Part 7.

20. Ibid.

21. L. K. Caldwell, "The Energy Crisis and Environmental Law: Paradox of Conflict and Reinforcement," *New York Law Forum*, January 1975, p. 50.

22. Northern Great Plains Resource Program, *Draft Report* (Denver, 1974), p. 1.

23. James G. Phillips, "Environmentalists, Utilities Argue Over East, West Mining," *National Journal Reports*, 6 July 1974, p. 1018.

24. Katharine Fletcher, ed. *A Scientific and Policy Review of the Draft Environmental Impact Statement for the Proposed Federal Coal Leasing Program* (Washington, D.C.: Institute of Ecology, 1974), p. 68.

25. 43 U.S.C.A. 1501 et seq. (suppl. 1974).

26. Malcolm F. Baldwin, *The Southwest Energy Complex: A Policy Evaluation* (Washington, D.C.: Conservation Foundation, 1973).

27. Ibid.

28. Council on Economic Priorities, *Leased and Lost*, p. 13. See also Alvin M. Josephy, "Murder of the Southwest," *Audubon* 73 (1971).

29. Baldwin, *Southwest Energy Complex*, p. 5.

30. Ibid.

31. U.S., Congress, Senate, Committee on Interior and Insular Affairs, *Environment Problems Associated with the Production of Electrical Power by Coal-Fired Plants in the Four Corners Region of the Southwest U.S.: Hearings on Senate Resolution 45*, 92d Cong., 1st sess. 1971, 7 parts.

32. Council on Economic Priorities, *Leased and Lost*, p. 13.
33. Helena Huntington Smith, "The Winning of the West," *Washington Post*, January 1975.
34. Northern Great Plains Resource Program, *National and Regional Energy*, Part 7.
35. U.S. Department of the Interior, *U.S. Energy through the Year 2000* (Washington, D.C.; Government Printing Office, 1972).
36. Josel Haveman and James G. Phillips, "Independence Blueprint Weighs Various Options," *National Journal Reports* 6 (1974):1637.
37. Project Independence Background Paper presented at the Washington, D.C., Energy Conference, 11–12 February 1974, p. 21.
38. Les Gapay, "Elusive Goal: U.S. Self-Sufficiency in Energy by 1980 Is Unlikely, Experts Say," *Wall Street Journal*, 22 July 1974.
39. Freeman, *Energy*, p. 237.
40. *The National Energy Plan* (Washington, D.C.: Government Printing Office, 1977). See also Richard Lamm, "What President Carter's Energy Plan Means to Colorado," speech delivered in Denver, 16 May 1977.
41. Smith, "The Winning of the West," p. 4.
42. SBNA *Environmental Report* Curr. Dev. (1 November 1974) pp. 1040–50.
43. Caldwell, "The Energy Crisis," pp. 53–54.
44. Box, *Rehabilitation Potential*.
45. Gapay, "Elusive Goal."
46. Fletcher, *Scientific and Policy Review*, p. 149.
47. Southwest Research and Information Center, "Gasification Plants Will Take Water from the Navajo Indian Irrigation Project" (Albuquerque, November 1974), p. 4. See also Ben A. Franklin, "Clash Over Water Delays, U.S. Gas Project in West," *New York Times*, 12 June 1975.
48. Fletcher, *Scientific and Policy Review*, p. 160.
49. Ibid., p. 165.
50. Ibid., pp. 159–73. See also U.S. Environmental Protection Agency, *Southwest Energy Study: Report on the Work Group on Air Pollution Aspects* (Washington, D.C.: U.S. Government Printing Office, 1972), p. 8; Albert G. Melcher, *Wanted—A Better Approach to Energy Planning* (Denver: Rocky Mountain Center on Environment, 1972), p. 10.
51. Alvin M. Josephy, "Agony of the Northern Plains," *Audubon*, July 1973.
52. Federation of Rocky Mountain States, *Energy Development*, Section 2.
53. Ibid., p. 25. See also Box, *Rehabilitation Potential*, p. 35.
54. Federation of Rocky Mountain States, *Energy Development*, p. 25.
55. Box, *Rehabilitation Potential*, p. 9.
56. Federation of Rocky Mountain States, *Energy Development*, p. 29.
57. Bill Strabala, "Forty Power Plants Mapped in State," *Denver Post*, 31 July 1977.
58. Author interview with Jim Monaghan, adviser to governor of Colorado, 10 March 1978.
59. "The Word From the West—Where It's At, Where It's Going," *National Journal Reports*, 25 February 1978, p. 309.

Chapter 3

1. John G. Jenkins, "$4.06 Short" (B.A. thesis, Princeton University, 1974), p. 8.
2. Ibid., p. 23.
3. Frank H. Jonas, "The Western Scene," in *Politics in the American West*, ed. Jonas (Salt Lake City: University of Utah Press, 1962), p. 2.
4. Walter Prescott Webb, *The Great Plains* (Boston: Ginn and Co., 1931).

5. Carl Frederick Kraenzel, *The Great Plains in Transition* (Norman: University of Oklahoma Press, 1955).

6. Frederick Jackson Turner, *The Frontier in American History* (New York: Henry Holt, 1920) and *The Significance of Sections in American History* (New York: Henry Holt, 1932).

7. John Walton Caughey, *The American West: Frontier and Region* (Los Angeles: Ward Ritchie Press, 1969), p. 18.

8. Ibid., p. 22.

9. Ibid., p. 23.

10. Ibid.

11. Edgar Lane, "The Political West: Section or State of Mind?" paper delivered to Western Political Science Association, 1971.

12. Ibid., p. 2.

13. Ira Sharkansky, *Regionalism In American Politics* (New York: Bobbs-Merrill, 1970).

14. Lane, "The Political West," p. 12.

15. Sharkansky, *Regionalism*, p. 138.

16. Neal A. Maxwell, "The West On Capitol Hill," in Jonas, *Politics*, p. 470.

17. Lane, "The Political West," p. 14.

18. K. Ross Toole, *The Rape of the Great Plains* (Boston: Little Brown, 1976). See also Charles G. Bell, "Politics in the West," *The Western Political Quarterly* 28 (1975): 237–39; L. K. Caldwell, "Synthesis," *Western Wildlands* 3 (1977): and James L. Plummer, "The Federal Role in Rocky Mountain Energy Development," *Natural Resources Journal*, April 1977.

19. Jonas, "The Western Scene," p. 2.

20. Ibid., p. 13.

21. Lane, "The Political West," p. 15.

22. Maxwell, "Capitol Hill," p. 22.

23. Caughey, *The Amerivan West*, p. 24.

24. Western Governors' Task Force on Regional Policy Management, *Regional Policy Management: The Task Ahead* (Denver, 1976).

25. Martha Derthick, *Between State and Nation* (Washington, D.C.: Brookings Institution, 1974), p. 132.

26. "Regional Commission To Aid 'Boom Towns,' " *Denver Post*, 15 December 1974.

27. "Governors To Decide on Regional Airport," *Independent Record*, Montana, 11 December 1974.

28. Derthick, *Between State and Nation*, p. 148.

29. Council of State Governments, *Interstate Compacts 1783–1970: A Compilation* (Lexington, 1971), p. 60.

30. Western States Water Council, *Western States Water Requirements for Energy Developments to 1990* (Salt Lake City, 1974), p. 40.

31. Maxwell, "Capitol Hill," pp. 491–92.

32. David Sumner, "Colorado: The Last Decade," *Colorado*, July–August 1975, p. 68.

33. Maxwell, "Capitol Hill," p. 493.

34. Jenkins, "$4.06," p. 152.

35. Dick Redburn, "Herschler Says Development Must Be on States' Terms," *Casper Star*, 22 October 1974.

36. Gary J. Wicks, "Statements on Project Independence," Montana Department of Natural Resources and Conservation, 7 August 1974, p. 25.

37. Lauren S. McKinsey and Louis D. Hayes, "Regional Energy Developments In Comparative Federal Systems" (University of Montana, April 1974), p. 21.

38. Daniel M. Hall, "Montana Coal: The Alamo of Interstate Commerce," *North Dakota Land Review* 51 (1974):439–56.

39. Author interview with Jack Neckels, former Director of Division of State Planning, North Dakota, 12 September 1975.

40. Governor Link of North Dakota to Dudley Favor, 13 August 1975, and Statement by Governor Link, Project Independence Hearings, Billings, Montana, 4 September 1974.

41. Author interview with C. Emerson Murray, North Dakota Regional Environmental Assessment Program, 30 October 1975.

42. "Castro Calls for Harmony," *Phoenix Independent Republic*, 7 January 1975.

43. Author interview with John McComb, Sierra Club Representative, Tucson, 14 September 1975.

44. Author interviews with Brant Calkin, Clearinghouse Director, Santa Fe, 15 September 1975 and 14 April 1978.

45. Interview with McComb, 14 September 1975.

46. Grace Lichtenstein, "Utah Wants Power Plan Despite Ecology Effect," *Tucson Star*, 9 August 1975.

47. Letter to author from Michael H. Annison, Executive Vice-President, Federation of Rocky Mountain States, 22 September 1975.

48. *Proceedings of the Ninth Annual Meeting of the Federation of Rocky Mountain States*, 9-10 October 1973 (Denver), p. 16.

49. *Proceedings of the Tenth Annual Meeting of the Federation of Rocky Mountain States*, 4-5 September 1974 (Denver), p. 12.

50. Federation of Rocky Mountain States, *Energy Development in the Rocky Mountain Region: Goals and Concerns* (Denver, 1975), p. 134.

51. Author interview with Edwin Zaidlitz, Bureau of Land Management, Billings, Mont., 18 September 1975.

52. Author interview with E. W. Christiansen, former Lieutenant Governor, Montana, 27 September 1975.

53. Author interview with Jack Neckels, former Director, North Dakota Division of State Planning, 12 September 1975.

54. Ralph A. Luken, *Economic and Social Impacts of Coal Development in the 1970's for Mercer County, North Dakota* (Washington, D.C.: 1974, Chapter 2.

55. Federal Energy Administration, *Project Independence Blueprint: Transcript of First Public Hearing, Denver 6-9 August 1974* (Washington, D.C.:Government Printing Office).

Chapter 4

1. Author interview with Governor Thomas L. Judge, 11 September 1975.

2. See David S. Broder, "Mountain West Allying against Energy Rape?" *Chicago Sun-Times*, 24 November 1974, and "Lamm Amazed at Unity on Energy Assertiveness," *Denver Post*, 20 November 1974.

3. Author interview with Christopher Farrand, 21 January 1975.

4. Ibid.

5. Federation of Rocky Mountain States, *Special Report*, April 1974.

6. Author interviews with Michael H. Annison, 18 September 1975, and Warren C. Wood, 19 September 1975.

7. Author interview with Wood.

8. Author interview with Stanley Womer, 26 September 1975.

9. "Nine-State Energy Policy to be Discussed by Democrats," *Missoulian*, 6 December 1974.

10. Statement sponsored by Thomas L. Judge, Governor of Montana, 7 December 1975.

11. This suspicion was expressed by some congressional delegations from the West to a confidential federal source.

12. This suspicion was expressed in confidence during interviews with certain state officials.

13. "Western Governors May Hire Sawhill," *Bozeman Chronicle*, 4 December 1975.

14. Robert Threlkeld, "Area Energy Post Hinted for Sawhill," *Rocky Mountain News*, 11 December 1974.

15. Minutes of meeting, Old West Regional Commission, Lincoln, Nebraska, 13 December 1974.

16. J. D. Holmes, "Sawhill Offers Opinions on Mountain Area Energy," *Superior Clipping Service* (Montana), 19 December 1974.

17. "An Area Energy Post Likely for Sawhill," *Rocky Mountain News*, 19 December 1974.

18. Sandra Hubbs, "Nine-State Coalition Planned to Protect Area Resources," *Cheyenne State Tribune*, 25 December 1975.

19. Author interview with William Armstrong, special assistant to Sawhill [now president of New York University], 2 October 1975.

20. "Sawhill Appointment Hit by Defense Fund," *Caldwell, Idaho, News-Tribune*, 25 December 1975.

21. "Sawhill Choice Challenged by Environmental Groups," *High Country News* (Lander, Wyoming), 3 January 1975.

22. Threlkeld, "Area Energy Post Hinted for Sawhill."

23. Jeff Stansbur and Edward Flattau, "Sawhill Loses Energy Post Bid," *Scottsdale Progress*, 11 February 1975.

24. Sam Cox, "Western States Meet to Discuss Crucial Energy Patterns in Santa Fe," *El Independiente* (Albuquerque), 25 July 1975.

25. Interview with Judge.

26. "Western Legislators Hear Resources Talk," *Billings Morning Gazette*, 28 January 1975.

27. Cox, "States to Discuss Energy Patterns." In addition, this information was shared in confidence in interviews in three different states and with three separate federal sources.

28. This information was shared in confidence by officials in the U.S. Department of Interior both in the region and in Washington, D.C., and officials of the two Title V Regional Commissions.

29. Letters to governors of eleven states from Rogers C. B. Morton, 3 January 1975.

30. Letters to governors of eleven states from William L. Rogers, 8 January 1975.

31. Letters to governors of ten states from Richard D. Lamm, 17 January 1975.

32. Letters to governors of eleven states from Rogers C. B. Morton, 17 January 1975.

33. Office of the White House Press Secretary, Message to the Congress, 15 January 1975, p. 12.

34. Dick Prouty, "U.S. Energy Needs Must Take Priority, Morton Tells West," *Denver Post*, 25 January 1975.

35. "Morton Challenges the West" (unidentified clipping), 28 January 1975.

36. Bob Ewegen, "Morton Talks to Governors on Tapping Western Coal," *Washington Post*, 25 January 1975.

37. Bob Ewegen, "Lamm Urges Energy Agency," *Denver Post*, 26 January 1975.

38. Prouty, "Energy Needs Take Priority."

39. Bureau of Land Management, State of North Dakota, *Memorandum of Understanding*, 22 January 1975.

40. Office of Ed Herschler, *Memorandum: Regional Energy Policy Office*, 28 January 1975.

41. "Four Western Governors: Energy Office Guides Set Up," *Denver Post,* 8 February 1975.
42. Sandra Hubbs, "Sawhill-Dominick Listed As Energy Office Administrator," *Denver Post,* 8 February 1975.
43. Draft of the Articles of Incorporation and Bylaws for a Non Profit Corporation, Memorandum from Office of Warren C. Wood, 11 February 1975, p. 12.
44. Author interview with Jack Neckles, former director of State Planning Division, 11 February 1975, p. 12.
45. Hubbs, "Sawhill-Dominick."
46. Memorandum from Chris Brown to New Mexico Governor Jerry Apodaca, 10 February 1975.
47. Daniel A. Chiszar, "Western Governors Form Coalition," Cheyenne *Eagle,* 8 February 1975.
48. Telegram from Governor of Wyoming, 8 February 1975.
49. Duane W. Bowler, "Good Place to Circle Wagons," *Mandan* [North Dakota] *Pioneer,* 29 January 1975.
50. Memorandum from Senator Frank Moss to members of the Western Conference," 10 February 1975.
51. "Congressman Reviving Western Conference," *Idaho Falls Evening Post Register,* 8 February 1975.
52. Draft Documents, Old West Regional Commission, 11 February 1975, p. 1.
53. Western Governors' Regional Energy Policy Office, Denver, 23 April 1975.
54. Minutes of meeting, Western Governors' Regional Energy Policy Office, 20 February 1975, pp. 48,49.
55. See Leonard Larsen, "Governors OK Energy-Policy Body," *Denver Post,* 22 February 1975, and "Western States Form Regional Energy Office," *Devils Lake* [North Dakota] *Journal,* 21 February 1975.
56. Statement of Concern, Western Governors' Regional Energy Policy Office, 20 February 1975.
57. Jerry Apodaca to Western Congressmen and Senators and Chairmen of Energy-Related Committees, 11 March 1975.
58. John F. Rhodes to Governor Apodaca, 19 March 1975.
59. Milton R. Young to Governor Apodaca, 19 March 1975.
60. Rocky Mountain Center on Environment, news release, 4 March 1975.
61. Jack O. Horton, "An Alliance of Interdependence," address to Western Governors' Conference on Agriculture, Billings, April 1975, p. 10.
62. Jerry Apodaca, Address to Western Governors' Conference on Agriculture, Billings, 1 April 1975.
63. Mel Reisner, "Water Continues to be Top Key to Westerns' Worry," *Miles City* [Montana] *Star,* 5 April 1975.
64. "Food As Weapon," *Helena Independent Record,* 10 April 1975.
65. Letter from Western Governors' Conference on Agriculture to President Ford, 1 April 1975, p. 3.
66. "Hathaway and Guy: How to Mind West?" *Idaho Falls Evening Post Register,* 10 April 1975.
67. Mel Reisner, "Not Nation's Energy Colony," *Wyoming News* [Worland], 8 April 1975.
68. Apodaca to Guy, 4 April 1975.
69. Memorandum from Ron Zee to Governor Apodaca, 14 April 1975.
70. Memorandum from Ron Zee to Governor Apodaca, 29 April 1975.
71. Interview with Wood.
72. Memorandum from William L. Guy to board of directors, 3 June 1975.

73. Minutes of special meeting, Western Governors' Regional Energy Policy Office, 8 June 1975.

74. See ibid. Governor Judge had testified before the U.S. Senate Committee on Interior and Insular Affairs regarding this act on 8 May 1975. See "Testimony of the Honorable Thomas L. Judge."

75. "A Pledge to Share Utah Fuel," *Deseret News* [Salt Lake City] 2 July 1975.

76. "The Viewpoint," *Pay Dirt* [Arizona], 28 April 1975.

77. "Energy: War Between the States," *Newsweek*, 5 May 1975.

78. Peter Katel, "At the Capitol," *Albuquerque Tribune*, 16 April 1975.

79. Letter from William L. Guy to all western governors, 4 June 1975.

80. Letter from William L. Guy to all Class B directors, 16 July 1975.

81. Records of the meeting of the WGREPO, 7 July 1975.

82. All details and quotations are derived from notes the author took in Santa Fe and from records of the meeting of the WGREPO, 27 July 1975.

83. All details and quotations are derived from notes the author took in Denver and from records of the meeting of the WGREPO, 29 July 1975.

Chapter 5

1. Arthur J. Magida and Richard Corrigan, "Trying to Pull the Plug on Uncle Sam's Energy Plans," *National Journal Reports*, 8 May 1976.

2. Robert W. Rafuse, Jr., *The New Regional Debate: A National Overview*, National Governors' Conference, April 1977, p. 3.

3. Magida and Corrigan, "Trying to Pull."

4. Linda E. Demkovich, "Governors' Departure Tests State Government Council," *National Journal Reports*, 27 September 1975.

5. Demkovich, "Governors' Departure."

6. Governors' Task Force for the Organization and Management of Multi-State and Regional Activities in the Mountain-Plains Area, *Task Force Briefing Paper*, 16 September 1976, appendices, p. 60.

7. Calvin Rampton to Representative Nancy Dick of Colorado, 18 March 1976.

8. Executive Office of the President, *Strengthening Public Management In The Intergovernmental System* (Washington, D.C.: Government Printing Office, 1975.)

9. Philip M. Burgess, "The State, The Region, and the Nation," address delivered to the Federation of Rocky Mountain States, October 1975.

10. Colorado State Planning Coordinating Council, "Policy Management and Executive Leadership in Colorado," 9 June 1976.

11. Ed Herschler to Calvin L. Rampton, 17 June 1976.

12. Author interview with Roy Romer, Treasurer of Colorado, 11 November 1977. Confirmed in interview with Philip Burgess, 10 March 1978. See also Richard Lamm, Rampton, 11 June 1976.

13. Larry McMurtry, "Sky Hunger," *Holiday Magazine* 58 (1977):39.

14. Task Force Minutes, 2 August 1976; author interview with Bruce Arkell, 1 May 1978.

15. Task Force Minutes, 25 August 1976, p. 3.

16. Ibid.

17. Task Force Minutes, 26 August 1976, attachment D.

18. Task Force, "Briefing Paper," 16 September 1976 p. 22.

19. Task Force, "Briefing Paper," 16 September 1976.

20. Ibid., chapters 3–5.

21. Dave Jonsson, "Regional Agencies: Any Value?" *Salt Lake City Tribune,* 22 June 1976.
22. Interview with Arkell.
23. Jonsson, "Regional Agencies."
24. Patrick Lamb, "States Investigating Government Groups," *Albuquerque Journal,* 28 September 1976.
25. Cy Ryan, "O'Callaghan Defends Governors' Conference," *Nevada State Journal,* 24 September 1976.
26. Ibid.
27. Lamb, "States Investigating."
28. Don Jepsen, "Multi-State Commissions Face Scrutiny," *Oregon Journal,* 13 November 1976.
29. A large literature on regional economic imbalances and regionalism has developed since about 1975. See *National Journal Reports,* 26 June 1976, and 27 November 1976, and 2 July 1977. See also *National Observer,* 2 October 1976; and Rafuse, *The New Regional Debate."*
30. *Kevin Philips, "Regionalism Holds Key to Future," Pensacola News Journal,* November 1976.
31. William L. Guy to Lamm, 23 August 1976.
32. Letters from Jay Hammond to sixteen Governors, 27 September 1976.
33. Task Force Minutes, 17 October 1976.
34. Jepson, "Multi-State Commissions."
35. Author interview with Jim Monaghan, Aide to Governor Lamm of Colorado, 10 March 1978.
36. Governors' Task Force on Regional Policy Management, *Regional Policy Management: The Task Ahead,* 10 December 1976, p. 107.
37. Todd Phipers, "Western Governors Wrestling Multi-State Groups, Pacts," *Denver Post,* 3 December 1976.
38. Letters from William L. Guy to Arthur Link and Richard Kneip, 17 December 1976.
39. Task Force Session and Plenary Session of the Governors, 10 December 1976, p. 6.
40. Neal R. Peirce, "States Making Their Presence Felt," *Denver Post,* 12 December 1976. See also idem, "A Strong Western Voice on Energy," *Denver Post,* 2 January 1977.
41. Neal R. Peirce. "Energy Coordination: The West Gets Its Act Together" (original draft), Neal Peirce Syndicated Column, 24 December 1976.
42. Jerry Hagtrom and Neal R. Peirce, "Western States Join in Forming a United Energy Front," *National Journal Reports,* 5 February 1977.
43. Task Force Minutes, 21 March 1977, p. 13.
44. "Energy Resources," background paper for meeting of governors, 10 December 1976.
45. Energy Interim Steering Committee, "Establishing a Verified Energy Resources Policy Office," 7 February 1978.
46. Task Force Meeting, 18 March 1976. p. 9.
47. Task force Meeting, 27 June 1977.
48. John L. Watson to author, 28 August 1978.
49. Task Force Meeting, 21 March, 1977. p. 6.
50. Task Force Meeting, 21 March 1977.
51. Jay Hammond to Raul Castro 25 May 1977.
52. Memorandum to Task Force Members from Roy Romer and Bill Gordon, 5 May 1977.
53. Task Force Meeting, 2 June 1977. The term "huddling" was used by Bruce Arkell of Nevada in an interview with the author.
54. See Norris's Report. Task Force Meeting, 2 June 1977.

55. Task Force Meeting, 2 June 1977.
56. Ibid.
57. Leonard Larsen, "Lamm Scathes Carter Stance on West: Resources Role Called Sacrificial," *Denver Post,* 10 July 1977.
58. Author interview with Russell Sadler, Oregon jounalist, 3 September, 1977.
59. Thomas Judge to John Evans, 16 June 1977.
60. Interview with Sadler.
61. Address by Cecil Andrus, Western Governors' Conference, 1 September, 1977.
62. Author interview with George Ariyoshi, Governor of Hawaii, 9 November 1977.

Chapter 6

1. "Western Resources: Growth's Exploitation," *Congressional Quarterly* 17 July 1978, pp. 1907–10
2. Author interview with John Watson, Executive Director, WIEB/WINB, April 10, 1978. See also ibid.
3. Task Force Minutes, "Briefing Paper," 16 September 1976, p. 43.
4. WESTPO, *Balanced Growth and Economic Development: A Western White Paper,* February 1978.
5. Fulmer Mood, "The Origin, Evolution and Application of the Sectional Concept, 1750-1900," in *Regionalism in America,* ed. Merrill Jensen (Madison: University of Wisconsin Press, 1951).
6. Rupert B. Vance, "The Regional Concept as a Tool For Social Research." in Jensen, ed. *Regionalism in America.*
7. Donald Davidson, "Regionalism and Sectionalism Meet," *Social Forces* 13 (1934):23–31.
8. Frederick Jackson Turner, *The Significance of Sections in American History.* (New York: Henry Holt, 1932), p. 288.
9. Howard W. Odum and Harry Estill Moore, *American Regionalism: A Cultural Historical Approach to National Integration,* (New York: Henry Holt, 1938).
10. Mood, "Orgin of Sectional Concept," p. 425.
11. Vance, "The Regional Concept," pp. 137–40.
12. Davidson, "Regionalism," p. 30.
13. Turner, *Sections in American History,* p. 288
14. Odum and Moore, *American Regionalism,* pp. 42–43.

Appendix

1. Felix Frankfurter, preface to *Regionalism in America,* ed. Merrill Jensen (Madison: University of Wisconsin Press, 1951), p. 425.
2. Fulmer Mood, "The Origin, Evolution, and Application of the Sectional Concept, 1750–1900," in Jensen, *Regionalism in America,* pp. 5–98.
3. Ibid., p. 32.
4. Ibid., pp. 32–33.
5. Ibid., p. 5.
6. Ibid., p. 87.
7. Advisory Commission on Intergovernmental Relations, *Multistate Regionalism* (Washington, D.C.: U.S. Government Printing Office, 1972), p. 5.
8. Lewis Mumford, *The Culture of Cities* (New York: Harcourt, Brace, 1938).

9. Howard W. Odum and Harry Estill Moore, *American Regionalism: A Cultural-Historical Approach to National Integration* (New York: Henry Holt, 1938), p. 693.
10. Rexford Tugwell, "Constitution for a United Republic of America," *The Center Magazine*, September 1970, pp. 24–49.
11. Rupert B. Vance, "The Regional Concept as a Tool for Social Research," in Jensen, *Regionalism in America*, pp. 137, 140.
12. Frederick Jackson Turner, *The Significance of Sections in American History* (New York: Henry Holt, 1932), p. 288.
13. Donald Davidson, "Where Regionalism and Sectionalism Meet," *Social Forces* 13(1934):23.
14. Odum and Moore, *American Regionalism*, p. 43.
15. Ibid., p. 18.
16. Walter Isard, "Regional Science, The Concept of Region, and Regional Structure," in *Papers and Proceedings of the Regional Science Association* (Cambridge, Mass., 1956), pp. 1–14.
17. Odum and Moore, *American Regionalism*.
18. National Resources Committee, *Regional Factors in National Planning* (Washington, D.C., 1935).
19. Ibid., p. 145.
20. Odum and Moore, *American Regionalism*, bibliography.
21. W. B. Munro, "Do We Need Regional Governments?" *Forum*, January 1928.
22. See Marshall E. Dimock, "Political and Administrative Aspects of Regional Planning," in *Proceedings of Joint Conference on Planning, 1936*, and Marshall E. Dimock and G. C. S. Benson, *Can Interstate Compacts Succeed?*, Public Policy Pamphlet no. 22 (Chicago: University of Chicago, 1937).
23. W. Y. Elliott, *The Need for Constitutional Reform* (New York: McGraw, 1935).
24. See their work in *Regional Factors in National Planning*.
25. J. W. Fesler in the *American Political Science Review* 30 (1936):257–68. and *Social Forces* 15 (1936) 12–21; and Donald Davidson in the *American Political Science Review* 30 (1936):257–68, *Social Forces* 13 (1934):23–31; and *The Annals* (January 1949), and *The Attack of Leviathan: Regionalism and Nationalism in the United States* (Chapel Hill: University of North Carolina Press, 1938).
26. Mumford, *Culture of Cities*.
27. David B. Truman, *Administrative Decentralization* (Chicago: University of Chicago Press, 1940).
28. Kurt Wilk, *Decentralizing Governmental Work* (New York: Institute of Public Administration, 1962).
29. W. Brooke Graves, *American Intergovernmental Relations* (New York: Charles Scribner's Sons, 1964).
30. Arthur Maass, ed., *Area and Power* (Glencoe, Ill.: Free Press, 1939).
31. James W. Fesler, *Area and Administration* (University: University of Alabama Press, 1949), p. 34.
32. Thomas C. Donnelly, ed. *Rocky Mountain Politics* (Albuquerque: University of New Mexico Press, 1940).
33. V. O. Key, *Southern Politics: In State and Nation* (New York: Alfred A. Knopf, 1949).
34. Vincent Ostrom, "The Political Dimension of Regional Analysis," *Papers and Proceedings of the Regional Science Association* (Cambridge, Mass., 1956), 2:86.
35. Key, *Southern Politics*.
36. Duane Lockard, *New England State Politics* (Princeton, N.J.: Princeton University Press, 1959).
37. John Fenton, *Politics in the Border States* (New Orleans: Hauser, 1956); idem, *Midwest Politics* (New York: Holt, Rinehart and Winston, 1966).

38. Frank Munger, *American State Politics: Readings for Comparative Analysis* (New York: Thomas Y. Crowell, 1966).

39. Frank H. Jonas, *Western Politics* (Salt Lake City: University of Utah Press, 1961); idem, ed., *Politics in the American West* (Salt Lake City: University of Utah Press, 1962).

40. Donnelly, *Rocky Mountain Politics.*

41. Daniel J. Elazar, *American Federalism: A View from the States* (New York: Thomas Y. Crowell, 1966).

42. Ira Sharkansky, *Regionalism in American Politics* (New York: Bobbs-Merrill, 1970), p. 179.

43. Lockard, *New England State Politics,* p. 7.

44. Donnelly, *Rocky Mountain Politics,* p. 10.

45. Jonas, *Politics in the American West.*

46. Ibid., p. 2.

47. Munger, *American State Politics.*

48. Elazar, *American Federalism.*

49. Sidney Tarrow, "Local Constraints on Regional Reform," *Comparative Politics* 7(1974):1–36.

50. Joseph G. M. Hilhorst, "Spatial Structure and Decision Making," in *Issues in Regional Planning,* ed. J. G. M. Hilhorst and David M. Dunham (The Hague: The Hague Institute of Social Studies, 1971), pp. 47–72.

51. Sharkansky, *Regionalism in American Politics,* p. 5.

52. Ibid., p. 124.

53. Ibid., p. 144.

54. Jack L. Walker, "The Adoption of Innovations by the American States," cited ibid., p. 11.

55. David K. Hartley, *An Assessment of Regional Growth Issues in the United States* (Columbus, O.: Academy for Contemporary Problems, 1974), p. 145.

56. Richard H. Leach, *American Federalism* (New York: W. W. Norton and Co., 1970), p. 232.

57. Vincent V. Thursby, *Interstate Cooperation* (Washington, D.C.: Public Affairs Press, 1953), pp. 10–11.

58. Martha Derthick, *Between State and Nation: Regional Organizations of the United States* (Washington, D.C.: Brookings Institution, 1974), p. 242.

59. Marian E. Ridgeway, *Interstate Compacts: A Question of Federalism* (Carbondale: Southern Illinois University Press, 1971), p. 292.

60. Charles McKinley, "The Management of Water Resources Under the American Federal System," in *Federalism: Mature and Emergent,* ed. Arthur W. MacMahon (New York: Doubleday and Co., 1955), p. 347.

61. See ibid., pp. 346–47.

62. Richard H. Leach and Redding Sugg, Jr., *The Administration of Interstate Compacts* (Baton Rouge: Louisiana State University Press, 1959), p. 216.

63. Richard H. Leach, "The Interstate Oil Compact: A Study of Success," *Oklahoma Law Review* 10(1957):274

64. Derthick, *Between State and Nation.*

65. U.S. Advisory Commission on Intergovernmental Relations, *Multistate Regionalism, A Commission Report* (Washington, D.C., 1972), p. 271.

66. Weldon V. Barton, *Interstate Compacts in the Political Process* (Chapel Hill: University of North Carolina Press, 1965), p. 197.

67. Joan B. Aron, *The Quest for Regional Cooperation* (Berkeley and Los Angeles: University of Californa Press, 1969), p. 225.

68. Robert Warren and Geoffrey Wandesforde-Smith, *A Comparative Analysis of American and Canadian Governmental Arrangements for the Development of Regional*

Water Policy in the Columbia River Basin (Seattle: University of Washington, Department of Political Science, 1970).

69. Roscoe C. Martin, Guthrie S. Burkhead, Jesse Burkhead, and Frank J. Munger, *River Basin Administration and the Delaware* (Syracuse: Syracuse University Press, 1960).

Index

Advisory Commission on Intergovernmental Relations, 214
Alaska, energy problems of, 40–41
Aldrich, Harold, 24
Andrus, Cecil B., 80, 83, 85, 86, 89, 93, 136, 144, 156, 157
Annison, Michael H., 79
Apodaca, Jerry, 4, 69, 76, 80, 90, 95–106 passim, 116, 118, 119, 127, 137, 144
Appalachian Regional Commission, 211, 213, 214
Ariyoshi, George, 156, 159, 161
Ariyoshi Compromise, 159
Arizona: energy problems of, 41–42; energy-related legislation of, 68–69
Arkell, Bruce, 132, 133, 136, 140, 150
Aron, Joan B., 215, 216
Aspinall, Wayne, 64
Atlantic States Marine Fisheries Commission, 215

Babbit, Bruce, 180
Balanced Growth and Economic Development: A Western White Paper (WESTPO), 181
Barton, Weldon V., 210, 215
Blueprint (for Project Independence), 25, 26, 27
Briggs, Kent, 148
Bucks, Dan, 109, 110, 113, 150
Burgess, Philip, 126, 127, 128, 131, 136, 139, 141, 145, 146, 147, 149, 151, 154, 162, 178
Burns, Keith, 136, 140

Caldwell, L. K., 6, 56
Campbell, Jack, 178
Carey, Hugh, 124
Carlson, Jack, 116, 117, 118
Carter, Jimmy, 19, 40, 175. *See also* National Energy Plan
Carter administration, 151, 155, 160
Carter Plan, 19, 28–30, 39
Castro, Raul H., 68, 79, 80, 96, 108, 114, 115, 119, 120, 144
Caughey, John Walton, 53, 54, 55, 58
Central Arizona Project, 21
Charlton, Tom, 150
Christiansen, Bill, 3, 66
Clark, Noel, 109, 149
Class B directors, 106, 107, 109, 113, 114, 115

coal, 13, 14, 15, 20–21, 22, 25–26, 27. *See also* energy resources
Coalition of Northeastern Governors (CONEG), 124, 137
Colorado: energy problems of, 42–43; Policy Management Council of, 127
Colorado River Compact, 61
CONEG, 124, 137
Conference of Western Democrats, 62
Conference of Western Senators, 62
cooperative agreements (in Title XXX, Code of Federal Regulations), 30, 174
Council of State Governments (CSG), 60, 61, 210; NGC and, 124; Task Force and, 135, 136; WGC and, 133, 139, 153
Cox, Archibald, 85
Craigie, William, 193
CSG. *See* Council of State Governments
Culture of Cities, The (Mumford), 199
Cunningham, Donald, 100

DATF, 151, 152, 156
Davidson, Donald, 187, 188, 196, 199
Delaware River Basin Commission, 211, 213
Democratic Governors' Conference (Hilton Head, S.C.), 3, 76, 77, 78, 84
Dent, Frederick B., 95
Denver Research Institute, 100
Derthick, Martha, 60, 210, 213
Desmond, Humphrey, 194
Dimock, Marshall E., 199
Domenici, Pete V., 99
Dominick, David, 91, 99
Donnelly, Thomas C., 200, 203, 204

Education Commission of the States, 137
Elazar, Daniel J., 203, 205, 207, 208
Elliott, William Y., 199
EMARS, 30, 116, 118, 174
energy facility expansion, 38–39
Energy Minerals Allocation Recommendation System (EMARS), 30, 116, 118, 174
Energy Policy Project (Ford Foundation), 27
energy problems, by state, 37–50
energy resources: development of and environmental impact, 31–36; development of, rationale for, 20–21; distribution of, 14; extent of, 12–19. *See also* coal; geothermal energy; natural gas; oil; uranium
Energy Resources Policy Office, 144, 148. *See also* Western Interstate Energy Board
Energy Steering Committee (of the Task Force), 150
Engdahl, David, 149
Environmental Defense Fund, 84
Environmental Policy Center, 84, 85
Evans, Dan, 136, 146, 155
Exon, J. James, 83, 90, 95, 96, 103, 115

Fannin, Paul, 99
Farrand, Christopher, 78
Federal Coal Leasing Amendments Act (1976), 30, 174
Federal Mineral Leasing Act (1920), 108
Federal Regional Councils, 133, 195
Federal Strip Mine Reclamation Act, 80
Federation of Rocky Mountain States, 60, 63, 73, 100, 187, 210; description of, 61–62; energy development policy and, 71–72, 78–79; founding of WGREPO and, 82, 91, 125–26; Task Force and, 132, 134, 141, 147
Fenton, John, 203, 208
Fesler, James W., 199, 200
Ford, Gerald R., 40, 73, 80, 84, 86, 87, 88, 92, 101, 103, 108, 174, 175
Ford Administration, 30, 124
Four Corners Regional Commission, 60, 63, 107; description of, 59; founding of WGREPO and, 79–80, 82, 83, 90, 91, 93, 95, 99, 106; regionalism and, 73; Task Force and, 128, 129, 132, 139
Frankfurter, Felix, 192
Franklin, Nick, 132, 140, 148
Freeman, S. David, 84
Freese, Richard, 99
Freudenthal, David, 110, 132, 140, 141, 149, 150

Gagnon, George, 142
Garn, Jake, 92
Garry, Dan, 141, 145
Gaus, John, 199
Geopolitik, 197, 199
geothermal energy, 13, 14, 18–19. *See also* energy resources
Geothermal Energy Research, Development and Demonstration Act (1974), 19
Gordon, Bill, 148, 150, 151, 153
Great Lakes Commission, 215
Gross, Fred, 149
Guy, William L., 4, 96, 99, 100, 104–15 passim, 118, 139, 144, 161–62, 175, 176, 177

Hall, Bob, 148, 150
Hammond, Jay, 139, 151, 153, 160, 168, 169, 180

Hansen, Clifford P., 92
Hart, Gary, 99
Hartley, David K., 209, 210
Harvey, D. Michael, 85–86, 99
Hathaway, Bob, 132, 141, 148
Hathaway, Stanley, 65, 83, 109, 111, 116
Hawaii, energy problems of, 43
Herschler, Ed, 4, 49, 65, 76, 84, 89, 90, 91, 92, 96, 115, 117, 118, 127, 128, 145, 157, 159, 180, 182
Hilhorst, Joseph G. M., 205–6
Hodsoll, Frank S. M., 94
Horton, Jack O., 78, 100–104 passim, 116, 117
hydroelectric power, 27. *See also* energy resources

Institute for Policy Research, 152
Institute of Ecology, 21, 34
Intergovernmental Science and Public Technology Program (National Science Foundation), 123
Interim Steering Committee (ISC) (of the Task Force), 148, 149
Interstate Oil Compact, 213
Interstate Oil Compact Commission, 215
Isard, Walter, 197
ISC, 148, 149

Jackson, Henry M., 99
Jenkins, John G., 52, 53
Jonas, Frank H. 53, 57, 203, 204, 207, 208
Judge, Thomas L., 3, 40, 66, 76–86 passim, 89, 90, 95, 96, 97, 98, 102, 103, 104, 107, 108, 116, 117, 119, 127, 128, 132, 142, 145, 149, 151, 153, 155–60 passim, 169, 178, 181, 182

Kash, Don, 6
Key, V. O., 200, 202, 203
King, Bruce, 69
King, Jason, 155
Kneip, Richard F., 83, 89, 95, 96, 107, 108, 115, 118, 119, 131–32, 139, 144, 145
Kraenzel, Carl Frederick, 53
Kramer, Bud, 155

Lamm, Richard D., 3, 13, 40, 64, 75–91 passim, 95–99 passim, 104, 105, 106, 107, 109, 114, 117, 119, 124, 126, 127, 128, 131, 135, 136, 139, 141, 142, 151, 152, 155–60 passim, 162, 169, 176, 178, 180
Lane, Edgar, 55, 56, 57
Leach, Richard H., 209, 212–13
League of Women Voters, 164
Link, Arthur, 4, 67, 83, 89, 90, 95, 96, 105, 107, 118, 139, 144, 146, 153, 180
Lockard, Duane, 203–4, 208
Love, John, 63
Lyons, William W., 78, 116, 117, 118

Maass, Arthur, 199
McComb, John, 69
McGee, Gale, 65
McKay, Gunn, 92
McKinley, Charles, 211–12
McMurtry, Larry, 130
Mallicoat, Dale, 149, 150
Martin, Roscoe, 215, 216
Matheson, Scott, 142, 156, 159, 180
Maxwell, Neal A., 56, 57, 62, 208
Melcher, John, 92, 99
Merriam, Charles, 199
Metcalf, Lee, 99
Metropolitan Regional Council (New York), 215, 216
Metzger, Charles, 109, 110, 111, 114, 149
Michel, Mark, 99
Minier, Alan B., 78, 91, 93, 94, 95, 106
Missouri River Basin Commission, 59
Monaghan, Jim, 39, 40, 109, 110, 111, 114, 115, 128, 133, 139, 140, 141, 144, 148, 149, 151
Montana: energy problems of, 44; energy-related legislation of, 66
Mood, Fulmer, 187
Moore, Harry Estill, 187, 189, 195–99 passim, 214
Morse, Jedediah, 194
Morton, Rogers C. B., 23, 29, 78, 86–90 passim, 94, 95, 157
Moss, Frank E., 62, 92, 99, 176
Mountain-Plains Task Force. *See* Task Force
MSOs (multistate organizations), 178, 186; Federation study of, 125–26; national executive branch, 124, 125; Task Force and, 128, 131, 133–39; in the West, 5, 58–63, 70–74; WESTPO and, 158–59, 182, 183, 185. *See also* regional organizations, studies of
multistate organizations. *See* MSOs
Mumford, Lewis, 195, 199
Munger, Frank, 203, 204–5, 208
Munro, William B., 199

Nagel, Joe, 140
National Academy of Engineering, 27
National Academy of Science, 36
National Democratic Charter Convention (1974), 80, 81
National Energy Council, 86

National Energy Plan (Carter), 19, 28–30, 39
National Energy Supply Strategy, 29
National Environmental Policy Act (1969), 30, 87, 117
National Governors' Association, 186. *See also* National Governors' Conference
National Governors' Conference (NGC), 144, 176; founding of WESTPO and, 128; MSOs and, 124–25, 128, 134, 186; in New Orleans, 106, 107; in Washington, D.C., 91, 149
National Resources Committee, 198, 199
National Science Foundation, 68, 123
natural gas, 13, 14, 17–18, 25, 27. *See also* energy resources
Natural Resources Planning Board, 195
Navajo Indian Irrigation Project, 33
Navajo Project Participation Agreement, 21–22
Nelson, Bruce, 132, 148
Nevada, energy problems of, 44–45
New England Congressional Caucus, 137
New Mexico: energy problems of, 45–46; energy-related legislation of, 69
NGC. *See* National Governors' Conference
NGPRP (Northern Great Plains Resource Program), 66, 170, 187; findings of, 25–26, 33, 36; founding of, 25, 72–73; founding of WGREPO and, 91, 100, 101, 104
Nixon, Richard M., 26, 27, 28, 40
Nixon administration, 30, 85, 167
Norris, Jerry, 136, 153, 154, 155
North Central Power Complex, 23–26, 33
North Central Power Study, 19, 20, 21, 25, 29, 30, 65, 66, 73
North Dakota: energy problems of, 46–47; Regional Environmental Assessment Program of, 68
Northeast-Midwest Economic Advancement Coalition, 137
Northern Great Plains Resource Program. *See* NGPRP
Northern Plains Resource Council, 84–85
Northwest Energy Policy Project, 147
Northwest Energy Policy Study, 150
nuclear power, 27. *See also* uranium; energy resources

O'Callaghan, Mike, 80, 107, 108, 120, 131, 136, 142, 145, 158, 159, 160, 184
Odum, Howard W., 187, 189, 195–99 passim, 214
Office of Policy Management (of WEPC), 147, 150, 151, 152, 179
oil, 13, 14, 16, 17, 18, 27. *See also* energy resources
oil shale, 14, 16, 27, 28. *See also* energy resources
Oil Shale Leasing Program, 19, 20, 29; description of, 28
Old West Regional Commission, 60, 63, 67, 68, 85, 86, 107, 187; description of, 59; founding of WESTPO and, 128, 129, 132, 139; founding of WGREPO and, 77, 80, 82, 83–84, 90, 91, 93, 94, 95, 99, 106; role of, 73
Ono, Susumo, 140, 141
Ostrom, Vincent, 198, 200, 201, 202, 207, 208, 209, 217

Pacific Northwest Regional Commission, 83, 90, 93, 139
Peters, Bill, 110, 112, 149–50
petroleum resources. *See* oil
Pierce, Neal R., 146
Pine, Patrick, 141, 147
Plummer, James L., 56
political geography (Geopolitik), 197, 199
Powell, John Wesley, 194
Project Independence, 19, 20, 21, 25, 29, 67, 70; description of, 26–27

Quincy, Josiah, 187, 193

Rampton, Calvin L., 70, 80, 96, 97, 101, 108, 115, 124–28 passim, 130, 131, 132, 135, 136, 139, 141, 142, 145, 146, 151, 153, 162, 169, 178
RAND Corporation, 146
Ray, Dixie Lee, 142, 153, 156, 157–58, 159
Regional Coordinating Council (of WEPC), 147
Regional Energy Advisory Board (of WESTPO), 182
Regional Energy Policy Office (of WGREPO), 91
regionalism, 7, 8, 53; as analytical concept, 192–97; major findings concerning, 217–18; in political science, 198–202; and politics, 202–9; in the West, 169–73, 185–89. *See also* sectionalism; West, the
Regionalism in American Politics (Sharkansky), 206
regional organizations, studies of, 209–16. *See also* MSOs
Regional Policy Management: A Framework for Decision (Task Force), 134
Regional Policy Management: The Task Ahead (Task Force), 138, 142, 153
Rhodes, John F., 99
Ridgeway, Marian E., 211

Rockefeller Foundation, 123
Rocky Mountain Center on Environment, 100
Rocky Mountain Institute for Policy Research (RMIPR), 129, 142, 159
Rogers, William L., 87
Romer, Roy, 128, 131, 134, 139, 140, 141, 147, 149, 151, 158, 159
Roncalio, Teno, 102
Ruckelshaus, William D., 85, 86

Sawhill, John C., 75, 76, 77, 78, 81–86 passim, 91, 116
sectionalism, 7, 53, 187–89, 205; as analytical concept, 192–96. *See also* regionalism; West, the
Sharkansky, Ira, 56, 203, 204, 206, 207, 208
Sirotkin, Phillip, 135
Smith, Chris, 155
Smith, James R., 24
Solar Energy Research Institute, 29
South Dakota, energy problems of, 47–48
Southern Growth Policies Board, 210
Southwest Energy Complex, 19, 20, 69, 73; description of, 21–23
Southwest Energy Study, 23, 29, 30
Southwest Power Complex. *See* Southwest Energy Complex
Staff Advisory Committee (of WGC), 153, 154
Straub, Robert, 135, 150, 153, 158, 159
Sugg, Redding, 212–13
Surface Mining and Control Act (1977), 175

Tarrow, Sidney, 205
Task Force (Mountain-Plains Task Force; Western Governors' Task Force on Regional Policy Management), 58, 178, 179; Interim Steering Committee of, 148, 149; role in founding of WESTPO, 130–57. *See also* Western Regional Drought Action Task Force
Tennessee Valley Authority (TVA), 22, 195, 199, 211, 213
Thursby, Vincent V., 212
Title V commissions, 133, 184, 211, 214
Title XXX, Code of Federal Regulations, 30, 174
Title II commissions, 133, 211, 214
Toole, K. Ross, 56
Touche-Ross and Company, 141
Truman, David B., 199
Tugwell, Rexford, 195
Turi, Andrew, 149
Turner, Frederick Jackson, 53, 187, 188, 189, 194, 195

TVA, 22, 195, 199, 211, 213

Udall, Morris K., 62, 92
U.S. Atomic Energy Commission, 27, 73
U.S. Bureau of Indian Affairs, 22
U.S. Bureau of Land Management (BLM), 22, 30, 72, 115; agreements between states and, 67, 68, 89, 101
U.S. Bureau of Mines, 116
U.S. Bureau of Reclamation, 20–24 passim, 33
U.S. Department of Agriculture, 72
U.S. Department of Commerce, 78, 94, 106, 177
U.S. Department of Energy, 29, 39, 182
U.S. Department of the Interior, 94, 175; development programs and, 20–30 passim; founding of WESTPO and, 156, 157; founding of WGREPO and, 78, 83, 88, 100–101, 106, 109, 111, 112, 115, 116; NGPRP and, 72
U.S. Energy Research and Development Administration, 29, 112, 117
U.S. Environmental Protection Agency, 35, 72, 85, 91
U.S. Federal Energy Administration, 29, 67, 74, 76, 77, 84, 116; founding of WGREPO and, 94, 106
U.S. Office of Management and Budget, 126
uranium, 13, 14, 17. *See also* energy resources
Utah, energy problems of, 48–49

Vance, Rupert B., 187, 188, 195, 196
VanderWalker, John, 73, 100, 104

Wandesforde-Smith, Geoffrey, 215, 216
Warnock, James, 109, 114, 150
Warren, Robert, 215, 216
Water Resources Planning Act (1961), 59–60
Water Resources Policy Office (of WESTPO), 151
Watson, John, 149
Webb, Walter Prescott, 53
WEPC (Western Governors' Policy Council), 147, 148, 149; founding of, 142, 144, 145, 146; Office of Policy Management of, 147, 150, 151, 152, 179; organization of, 143. *See also* WESTPO
West, the: defined, 7; as region, 52–58; responses to economic growth by, 63–70. *See also* regionalism; sectionalism
WEST, 22
Western Conference, 62

Western Congressional Caucus, 184
Western Energy Supply and Transmission Association (WEST), 22
Western Federation of Human Services, 179
Western Governors' Conference (WGC), 127, 178, 182; in Billings, Mont., 4, 92, 100, 104; description of, 61; in Jackson Hole, Wyo., 130, 133, 135; Task Force and, 131, 136, 138, 139, 142, 144, 147, 150, 153–59 passim, 162, 184
Western Governors' Policy Council. *See* WEPC
Western Governors' Policy Office. *See* WESTPO
Western Governors' Regional Energy Policy Office. *See* WGREPO
Western Governors' Task Force on Regional Policy Management. *See* Task Force
Western Interstate Commission on Higher Education (WICHE), 131, 135, 137, 181
Western Interstate Corrections Compact, 61
Western Interstate Energy Board, 147, 175
Western Interstate Energy Board/Western Interstate Nuclear Board, 150, 175, 179, 182
Western Interstate Nuclear Board. *See* WINB
Western Interstate Nuclear Compact, 148
Western Regional Drought Action Task Force (DATF), 151, 152, 156
Western Regional Republican Conference, 62
Western States' Land Commissioners, 129
Western States' Water Council, 61, 152, 179
WESTPO (Western Governors' Policy Office), 5, 6, 7, 8, 9, 124; founding of, 123, 129–62; organization of, 180; purpose of, 177–85; Regional Coordinating Council of, 182; regionalism and, 185–89 passim, 192, 217, 218; results of interviews concerning, 164–73; Water Resources Policy Office of, 151, WGREPO and, 173, 178, 179. *See also* Task Force; WEPC, WGREPO
WGC. *See* Western Governors' Conference
WGREPO (Western Governors' Regional Energy Policy Office), 4, 5, 6, 8, 9, 37, 40, 58, 127; founding of, 75, 81, 90–121; founding of WESTPO and, 123, 124, 125, 129, 154, 155, 156, 157, 160, 161, 162; Regional Energy Policy Office of, 91; regionalism and, 71, 183–89 passim, 192, 210, 217, 218; results of interviews concerning, 164–73; role of, 173–77; Statement of Concern of, 97–98, 99, 103; Task Force and, 130–34 passim, 137–49 passim; WESTPO and, 173, 178, 179. *See also* WESTPO
White House Conference on Balanced Growth (1978), 181, 182
WICHE, 131, 135, 137, 181
WIEB, 147, 175. *See also* Energy Resources Policy Office; WIEB/WINB
WIEB/WINB (Western Interstate Board/Western Interstate Nuclear Board), 150, 175, 179, 182
Wilk, Kurt, 199
WINB (Western Interstate Nuclear Board), 184; founding of WESTPO and, 155, 159; founding of WIEB and, 147, 150; Task Force and, 134, 139, 141, 144, 145, 148, 150, 153. *See also* WIEB/WINB
Womer, Stanley, 79, 80, 95
Wood, Warren C., 77, 78, 79, 83, 91, 93, 94, 95, 105–6
Wyoming: energy problems of, 49–50; Industrial Siting Act of, 65; Land Use Planning Bill of, 65

York, Carl, 100
Young, Milton R., 99

Zaidlitz, Edwin, 89
Zarb, Frank, 116
Zee, Ron, 99, 105, 106, 111, 149